Subjects of the World

Darwin's Rhetoric and the Study of Agency in Nature

Paul Sheldon Davies

16
EasyRead Large

Copyright Page from the Original Book

PAUL SHELDON DAVIES lives in Newport News, Virginia, with his wife and daughter. His first book, *Norms of Nature: Naturalism and the Nature of Functions*, was published in 2001. He teaches philosophy at the College of William and Mary.

The University of Chicago Press, Chicago 60637
The University of Chicago Press, Ltd., London

Printed in the United States of America

18 17 16 15 14 13 12 11 10 09 1 2 3 4 5
ISBN-13: 978-0-226-13762-9 (cloth)
ISBN-10: 0-226-13762-7 (cloth)

Library of Congress Cataloging-in-Publication Data

Davies, Paul Sheldon.
Subjects of the world : Darwin's rhetoric and the study of agency in nature / Paul Sheldon Davies.
p. cm.
Includes bibliographical references and index.
ISBN-13: 978-0-226-13762-9 (cloth : alk. paper)
ISBN-10: 0-226-13762-7 (cloth : alk. paper)
1. Teleology. 2. Philosophy of nature. 3. Evolution. 4. Agent (Philosophy) 5. Darwin, Charles, 1809–1882—Literary style. I. Title.
BD581.D25 2009
124—dc22

2008042033

∞ The paper used in this publication meets the minimum requirements of the American National Standard for Information Sciences—Permanence of Paper for Printed Library Materials, ANSI Z39.48-1992.

TABLE OF CONTENTS

FOR CASSONDRA CAILIN
MY DAUGHTER
SUBJECT OF THE WORLD
MORE THAN IMAGINATION
CAN EMBRACE
OR BEAR TO BEHOLD

1

A Progressive Orientation

Naturalism as Exploration

ONE

The Vividness of Truth: Darwin's Romantic Rhetoric and the Evolutionary Framework

A worthy naturalist, Humboldt thought, left no means "unemployed by which an animated picture of a distant zone, untraversed by ourselves, may be presented to the mind with all the vividness of truth, enabling us even to enjoy some portion of the pleasure derived from the immediate contact with nature." **ROBERT RICHARDS, *THE ROMANTIC CONCEPTION OF LIFE***

Darwin's Rhetoric

Charles Darwin was a shrewd rhetorician. The voice of *On the Origin of Species* has a conversational charm that is disarming and at times alluring. On occasion Darwin raises his pitch to sing the praises of living things in Romantically charged refrains. That too is rhetorically effective, for Darwin's defense of the theory of evolution by natural selection, while effectively strangling to death the argument from design, is expressed in tones that sometimes verge on reverence. The news that God is dead is put in the mouth not of a madman but of a man who retains at least some sensibilities of a traditional believer—a sensibility, above all, marked with an appreciation of the foibles and pretensions of human reason.

The contrast with Friedrich Nietzsche's rhetoric in *The Gay Science* is indeed stark. Nietzsche's messenger personifies the eminent collapse of our theological worldview. His derangement embodies the coming cultural crash; he is meant to disturb us by being so disturbed himself. Darwin's messenger aims at the opposite. He deliberately sets out to minimize for us the felt sense of agitation by adopting a tone that is mostly casual and sometimes pastoral. Usually Darwin is chatting with us or with other naturalists of his day, though occasionally he breaks into Romantic song. The rhetorical strategy is to adopt a tone and cadences that engender both calm and optimism—calm from the charm of conversation, optimism from the Romantic imagery and rhythms. The crucial assumption is that a poetically expressed appreciation of nature is likely to engender a felt sense of optimism, along with gratitude and even hope. The most urgent hope is that our love of nature and life—our love for what we formerly conceived as "God's creation"—can outlive our dying belief in God.

As a young naturalist, during his years aboard the *HMS Beagle* especially, Darwin was steeped in the works of Romantic naturalists, including Alexander von Humbolt. Far from proclaiming the death of our ancestral theological worldview, the German Romantics announce its transformation. God is no longer an external agent directing events from afar but is transmuted into something else, something suffusing all of reality. God is no longer outside the world; the world is now saturated with God. The font of life's fecundity, the wellspring of novelty and creative yearnings in nature, the mysterious inspiration of artistic souls—these are the embodiments of the Romantic God. For Humboldt, all of nature comprises a "harmoniously ordered whole"—he calls it the Cosmos—and the inspired soul not only grasps but becomes part of it. With echoes of Spinoza, Humboldt exclaims, "Everywhere the mind is penetrated by the same sense of grandeur and vast expanse of nature, revealing to the soul, by a mysterious inspiration, the existence of laws that regulate the forces of the universe." Precisely this is why, for Humboldt, the worthy naturalist, when describing some distant land or foreign species, is obliged to employ every available device to "animate" the scene for his reader. Such devices enable the reader to experience—to have one's mind "penetrated by"—and thus participate in the harmonious order of all things. It is the responsibility of the Romantic naturalist, according to Humboldt, to be an effective conduit of Spinoza's God, of "the universality and reciprocal limitation and unity of all the vital forces in nature."[1]

And that is why, on the Romantic view, the imagery and the rhythms with which naturalists speak are as vital as the substance of their theories; rhetoric serves a communicative function every bit as important as the propositions uttered. What I wish to suggest then is that we begin to appreciate Darwin's rhetorical insights by considering the question, What happens if we discard the theology latent in the Romantic view but retain the rhetorical strategies? Might we succeed in convincing those with theological instincts that the right view of life is decidedly nontheological? Might we make the evolutionary view of life palatable? Might we invite acceptance where Nietzsche's madman provokes resistance?

Darwin's Insights

Darwin's skill as a rhetorician does not diminish his accomplishments as a scientist. To the contrary, it confirms them, for his rhetorical insights are motivated by the general perspective provided by his view of life. Scientific inquiry is, of course, a form of human intercourse subject to non-rational as well as rational powers of persuasion and, as everyone knows, rhetorical devices are sometimes coercive and not merely persuasive. But beyond this truism there is the distinctively Darwinian point that we are evolved animals, products of earthly processes operating through long stretches in the history of this planet. And one thing we know in light of our animal history is that, like our close primate cousins and our more remote mammalian cousins, we are well equipped to anticipate and navigate our environment in ways that never reach conscious awareness or ways that rise to consciousness only after the fact. The architecture of our affective and low-level cognitive capacities is far more elaborate and pervasive than the architecture of our consciously accessible cognitive capacities. This makes plausible the suggestion that we, by virtue of our affective and cognitive capacities, are nonconscious but nevertheless vigilant surveyors and anticipators of the world. We sense threats before we become conscious of them; we are attracted to certain things before any conscious recognition of our desires; and so on. No wonder, then, the wisdom of the rhetorician. No wonder the care that Darwin, in crafting his public presentation, lavished upon the implicit, affective reactions that his theory was bound to provoke.

It is as if Darwin asked himself, "What are the most entrenched *habits of thought* and most entrenched *cognitive and affective dispositions* likely to inhibit a correct understanding of my theory?" As I shall construe them, the relevant habits of thought include the conceptual categories and other sorts of lore bequeathed to us by our cultural ancestors. They are the central ingredients of our conceptual scheme to which we are habituated by way of education and enculturation. We begin to appreciate the power of these habits as we acquire knowledge of our cultural history. And I shall construe the relevant cognitive and affective dispositions as those features of our psychology that inhibit or otherwise retard our best efforts at inquiry. These are capacities of our minds that sometimes interfere with capacities that motivate and guide our inquiring activities. We come to appreciate the depth and breadth of these infirmities as we make progress in neuroscience, psychology, and sociology.

Darwin knew that even his most educated peers would misunderstand him, so he sought to correct the habits or dispositions responsible for such errors. These are rhetorical strategies that add nothing of propositional content to the theory of evolution by natural selection; Darwin could have presented his view without them. But unless one's audience is responsive only to propositional contents—and that is *not* the kind of animal we are—a full and accurate presentation of one's theory must contain something more. It must include the resources to open up our affective and aesthetic sensibilities and the resources to bypass the biases of our psychological constitution. Darwin knew, after all, that he was offering a theory of life bound to disturb the worldview of nearly every educated person of his day. He knew he was offering theoretical dynamite to animals who, no matter the power of their intellect, are psychologically structured to detect and respond to threats even before they become consciously aware that they are being threatened.

It is here, in the study of our history and our constitution, in identifying the habits of thought and dispositions of mind that lead us astray, that Darwin's rhetorical insights have their greatest impact. Darwin's conversational charm and his occasional Romantic flourish are important, to be sure, but they are the relatively superficial devices with which he applied these insights. The more substantive application is evident in some of the explicit arguments he offers throughout his discussion. And in this book, I propose to let the more substantive applications of Darwin's rhetorical insights come to the fore. I shall do this by describing the ill effects of our history and our constitution on our attempts to study ourselves and by developing a variety of strategies for diminishing or reversing the retarding effects that these habits and dispositions sometimes have. My hope is that the strategies I offer are ones Darwin would have embraced if today's knowledge had been available to him.

If my account is compelling, we will have uncovered two important facts. First, the way that most contemporary theorists frame their intellectual tasks, especially those concerned with human nature, is in need of reform. It is a mistake to frame either our scientific or our philosophical inquiries in terms of conceptual categories that perpetuate habits of thought or dispositions of mind that conflict with the conclusions of our best sciences. And, as we will see, this point applies with considerable force to contemporary theorists who see themselves as robust naturalists; some of our very best naturalists are not naturalistic enough. Second, if my suggestions are on track, we will have discovered an orientation toward inquiry that helps us internalize the truth of evolutionary theory. It accomplishes this by integrating evolutionary theory into the very methods with which we frame our inquiries. In the introduction to his *Treatise on Human Nature,* David Hume expresses the intention of introducing the Newtonian methods of inquiry to the study of human nature. My intention is to do something analogous regarding Darwin's rhetorical insights. In good Humean fashion, we begin by trying to get a grip on the kind of inquirers we are—the historical and constitutional infirmities we bring to the table—and then formulating our intellectual tasks informed by that knowledge.

Nietzsche observed that important advances in human knowledge have been driven by advances in our methods of inquiry. On my reading of Darwin, his application of rhetoric illustrates Nietzsche's claim to great effect. It thus seems a good bet to try to develop further the devices Darwin employed so we may apply Nietzsche's insight to the intellectual problems of our own day.

Darwinian Strategies

It is easy enough to identify the crucial strategies employed by Darwin in the *Origin.* A few representative passages culled mainly from the first edition are all we need. Begin with the strategy Darwin employs in the course of trying to convince his readers that evolution occurs by means of artificial selection among domestic species. This is the central thesis of chapter 1, "Variation under Domestication." In later chapters, Darwin argues that evolution occurs in analogous fashion by means of selection in natural settings, but even here, in the context of artificial selection, he anticipates resistance from breeders to the hypothesis that current species descend from strikingly different ancestral species. His key insight is to anticipate that the intuitions and expectations of most plant and animal breeders are calibrated to the creationist hypothesis so widely accepted in his day. The key thought is that, since evolutionary theory undermines creationism, we must guard against the biasing effects that our creationist intuitions or expectations will no doubt generate:

One circumstance has struck me much; namely, that all the breeders of the various domestic animals and the cultivators of plants, with whom I have ever conversed, or whose treatises I have read, are firmly convinced that the several breeds to which each has attended, are descended from so many aboriginally distinct species [distinct species presumably created by god]. Ask, as I have asked, a celebrated raiser of Hereford cattle, whether his cattle might not have descended from long-horns, and he will laugh you to scorn. I have never met a pigeon, or poultry, or duck, or rabbit fancier, who was not fully convinced that each main breed was descended from a distinct species. (Darwin 1859, 28–29)

Having identified a ubiquitous form of resistance to the hypothesis of evolution, he immediately works to diffuse it by pointing to specific failures of imagination within the breeders' psychological fabric. Darwin does not ridicule breeders but points to the habits of thought or the affective and cognitive dispositions of mind that limit and color what they are inclined to see. And he explicitly extends the lesson to the context of natural selection:

The explanation, I think, is simple: from long-continued study they are strongly impressed with the differences between the several races; and though they well know that each race varies slightly, for they win their prizes by selecting such slight differences, yet they ignore all general arguments, and refuse to sum up in their minds slight differences accumulated during many successive generations. May not those naturalists who, knowing far less of the laws of inheritance than does the breeder, and knowing no more than he does of the intermediate links in the long lines of descent, yet admit that many of our domestic races have descended from the same parents—may they not learn a lesson of caution, when they deride the idea of species in a state of nature being lineal descendents of other species? (Darwin 1859, 34)

These may seem more or less pedestrian observations, but I think I can convince you otherwise. The suggestions that Darwin offers here and repeats in the concluding chapter of the *Origin* are, in my view, of considerable importance to a fruitful orientation toward inquiry. So let us have a closer look.

I think it is fair on interpretive grounds to say that here Darwin is offering two psychological speculations. I think it is also fair to read into both speculations an appeal to two basic elements, namely, biologically inherited dispositions of mind and culturally inherited habits of thought. The first speculation is that breeders are perceptually and affectively attuned to what is distinctive about their preferred breeds; and they *feel* that there is something unique to their own broods. This is a claim about the natural psychological dispositions of people who breed plants or animals. After all, you are hardly likely to engage in such activities without a prior love or at least curiosity for the organisms in your care. *Something* about these organisms—their potential for brightening your financial prospects, if nothing else—attracts and holds your interest and affections; if not, the requisite expenditures of time and money would surely drive you away. And, since something about the organisms predisposes you toward them, it is highly likely that these prior dispositions are reinforced through long hours of caregiving. Your ongoing cultivation only deepens your attachments, as gardeners and pet lovers will readily attest.

There is also a cultural element to this speculation, namely, that we are inclined to interpret our deeply felt attachments in culturally inherited terms, including some inherited concepts that conflict with evolutionary theory. We feel an attachment to Hereford, not to longhorns, and this fact engenders confidence that the difference cannot be one of degree. Yet part of the cause of our confidence is an antecedent sympathy toward the creationist view of the origins of life. Our feelings only serve to confirm what we are already inclined to believe, namely, that each species must have been uniquely crafted from above. Of course an atheist may, by temperament and years of caregiving, come to experience the same sense of uniqueness toward his preferred plants or animals, but a culturally sustained commitment to a creationist theology would surely accentuate the feeling. So both elements—the psychological dispositions and the habits of thought—help explain why breeders resist the suggestion that their preferred breeds are nothing more than well-marked varieties of some ancestral breed.

The further speculation is that, as the activity of breeding heightens our propensity to see our preferred organisms as unique, it simultaneously diminishes any inclination we might have toward adopting a deeply historical perspective on living things. We grow blind to the enormous power of small changes over thousands of generations; we refuse or simply do not think to "sum up" in our mind all the slight changes in traits accumulated over hundreds or thousands of successive generations. We fail, that is, to cultivate a *historical imagination;* we are taken in by the present moment because we are gripped with a felt sense of the uniqueness of this kind of organism. This failure is clearly encouraged by theologically based, culturally inherited habits of thought. Those who see the world through creationist eyes—those who assume that each species is a product of a special act of creation—feel no compulsion or encouragement to investigate the enormous formative powers of history. Why would they? On their view, the formative powers attributed to God are infinite, and the course of history on earth, no matter its duration, is impotent to alter what has been fixed by an infinite God.

These culturally inherited habits of thought are not the whole story, for Darwin apparently thought that the failure of historical imagination is a consequence, at least in part, of a failure to cultivate a sufficiently robust *psychological imagination* or, at minimum, a failure to employ our powers of reason to compensate for a limited psychological imagination. Consider once again the failure among breeders to sum up in their minds the accumulated effects of artificial selection. The problem here is twofold. There is the failure of the imagination to hold in a single embrace the alterations wrought over great stretches of time. But there is also the failure to notice in oneself this first failure of imagination, and this double-layered failure is what I am calling a failure of psychological imagination.

The failure of psychological imagination is quite general in scope. Indeed, the problem identified by Darwin in the reasoning of breeders is but one type of conflict engendered by the very constitution of our psychology. *Constitutional conflicts,* as I shall call them, are conflicts that arise when functionally distinct capacities of our minds are not in sync, when the outputs of one conflict with those of another. Darwin's appreciation of such conflicts is clearly on display in the well-known section of chapter VI, "Difficulties on Theory," where he responds to the objection that the theory of evolution by natural selection cannot account for "organs of extreme perfection":

To suppose that the eye, with all its inimitable contrivances for adjusting the focus to different distances, for admitting different amounts of light, and for the correction of spherical and chromatic aberration, could have been formed by natural selection, seems, I freely confess, absurd in the highest possible degree. Yet *reason tells me,* that if numerous gradations from a perfect and complex eye to one very imperfect and simple, each grade being useful to its possessor, can be shown to exist; if further, the eye does vary ever so slightly, and the variations be inherited, which is certainly the case; and if any variation or modification in the organ be ever useful to an animal under changing conditions of life, then the difficulty of believing that a perfect and complex eye could be formed by natural selection, though insuperable *by our imagination,* can hardly be considered real. (Darwin 1859, 186–87; my italics)

Darwin is insistent that although the argument from design has undeniable intuitive power, its power diminishes as we identify and correct both failures of imagination. Of particular interest is the claim in the final sentence concerning the contrast between reason and imagination. Our psychology, he claims, comprises different capacities that do not always operate in harmony and that sometimes conflict even in the course of scientific theorizing. The conflict in this case is not a simple clash between imagination and reason. The problem is that, first, our imaginations are limited in ways that our capacity for rational reflection is not and, second, we are typically oblivious to this crucial difference between imagination and reason. We tend not to notice that the felt confidence with which we assert that Hereford could not be descended from longhorns is a product of a relatively limited cognitive capacity. Furthermore, we tend not to notice the change in perspective wrought by the persistent application of reason to available evidence.

As we will see in later chapters, the most potent causal factor in our failures of psychological imagination is the pervasive effects of our nonconscious cognitive and affective capacities. We have only the sparest knowledge of what is going on in our own psychologies because reliable, conscious access to our own capacities is paltry. This is a central theme to which I shall return repeatedly. The second most potent causal factor is that we are blissfully unaware of just how unaware we are. This too is a point to which I will repeatedly return. And it appears to be an effect of our *naive realism,* our deep-seated sense that our perceptions of the world are direct and unbiased.[2] We thus are doubly vexed: a great deal of our mental lives occurs below the level of conscious awareness and, at the same time, we feel a false sense of confidence that all, most, or at least some of the really important parts of our mental lives are accessible to conscious awareness. Although the evidence for both these causal factors has been discovered only in recent years, I suspect that Darwin, had he known such evidence, would not have been overly surprised.

Indeed, that he was alive to such conflicts is particularly evident in his quoting with approval the following comments from Helmholtz concerning the "incongruities"—indeed, the "contradictions"—in the very structure of our perceptual capacities, including the workings of our eyes! Helmholtz says, "That which we have discovered in the way of inexactness and imperfection in the optical machine and in the image on the retina, is as nothing in comparison with the incongruities which we have just come across in the domain of the sensations. One might say that nature has taken delight in accumulating contradictions in order to remove all foundations from the theory of a pre-existing harmony between the external and internal worlds" (quoted in Darwin 1872b, 163).

The fact that Darwin explicitly endorses Helmholtz's view should grab our attention. For Helmholtz is adamant that the human mind is so riddled with contradictions—with accumulated constitutional conflicts—that it is no longer possible to seriously entertain the hypothesis that our immediate thoughts and feelings provide us with an accurate portrayal of the world. Darwin, it seems, is committed to the point about constitutional conflicts in no uncertain terms. To the extent we tend not to notice the accumulated contradictions in ourselves, we thereby fail to cultivate a psychological imagination. We ignore all general arguments and fail to sum up in our minds the range of conflicts likely to emerge from our very constitution.

This failure of psychological imagination, as I have said, helps sustain our lack of historical imagination. One rather striking way it does this is by encouraging us to feel that the theory of evolution by natural selection is indeed too limited to account for organs of extreme perfection. Here is how it works. We first behold the marvelous subtlety and complexity of the mammalian eye; this is a bit of reflection that is decidedly ahistorical. We then consider whether something so exquisite could have arisen out of the undirected or random causal-mechanical factors involved in natural selection; we ask ourselves this question without the painstaking employment of our reason that Darwin describes in response to this objection. And then, because we have failed to sum up the accumulated effects of long stretches of history, our aesthetic response to the dumb mechanisms of selection is no match for our felt appreciation of the apparent purposiveness with which the eye functions. We sense that the mechanisms of selection, no matter the length of time, are too paltry and stupid to ever produce something so clearly marked by intelligence and design. Thus we come to feel an undeniable gap between theory and our alleged observation. No wonder we find ourselves in a state of disbelief regarding evolutionary theory—such is the retarding power of an undeveloped psychological imagination.

And when we turn to Darwin's discussion of the struggle for existence, his appeal to the retarding effects of conflicts within our constitution is given full reign. The appeal to the failure of psychological imagination, to the distorting effects of our affective and cognitive dispositions, is impossible to miss: "We behold the face of nature bright with gladness, we often see superabundance of food; we do not see, or we forget, that the birds which are idly singing around us mostly live on insects or seeds, and are thus constantly destroying life; or we forget how largely these songsters, or their eggs, or their nestlings, are destroyed by birds and beasts of prey; we do not always bear in mind, that though food may be now superabundant, it is not so at all seasons of each recurring year" (Darwin 1859, 62).

This passage from chapter 3, "The Struggle for Existence," comes immediately after a dire warning that the nature of life will be misunderstood if we fail to recognize the enormous violence and death involved in the relentless creation of new life: "Nothing is easier than to admit in words the truth of the universal struggle for life, or more difficult—at least I have found it so—than constantly to bear this conclusion in mind. Yet unless it be thoroughly engrained in the mind, I am convinced that the whole economy of nature, with every fact on distribution, rarity, abundance, extinction, and variation, will be dimly seen or quite misunderstood" (Darwin 1859, 62). The universal struggle for life results from the fact that all living things are driven to reproduce at a rate that much exceeds mere replacement. Since nearly all of life is so driven, since renewable resources required for life are finite, and since the more or less constant conditions of life are sometimes lethal, destruction due to various forms of competition is inevitable. The relentless drive to produce more life in a world of finite resources and hostile conditions is, of necessity, a deeply destructive drive.

Although our culturally inherited habits of thought no doubt play a role here, our affective dispositions are most salient. Darwin is surely right that an intellectual grasp of the struggle for existence is relatively easy and that having a stomach for the actual facts involved is much harder. This is to highlight a powerful failure of psychological imagination. We grasp the struggle for existence at a level of comprehension relatively detached from our affective or visceral reactions; our capacity for reason processes the proposition without our imagination processing the brutal facts of the matter. So the failure of psychological imagination is twofold. There is a failure of integration between comprehension and imagination, and there is a failure to notice the failure of integration. The net result is that, because of such conflicts in the constitution of our psychology, much of the time we are oblivious to the mismatch between our capacities.

The real challenge then is to cultivate a more robust psychological imagination. The challenge in this case, in fully understanding "the whole economy of nature," is to internalize the destructiveness of life by "engraining" it in our minds. The challenge is to calibrate our intuitions, our aesthetic reactions, and our expectations in light of the struggle for existence. Unless we develop the habit of not merely seeing but also feeling all of life as pervasively destructive, we will continue to approach living things with hunches and gut feelings that limit or distort what we see. A failure of psychological imagination will lead us astray. We will fail to grasp the nature of living things.

The Source of Darwin's Strategies: The Romantic Naturalists?

The influence of the Romantic naturalists on Darwin is the topic of a wonderful recent discussion by Robert Richards. In *The Romantic Conception of Life,* Richards traces the development of biological theorizing from the early Romantic period in Germany, beginning with the works and lives of the brothers Schlegel (Friedrich and Wilhelm), Fichte, Schleiermacher, and others through those of Schiller, Schelling, von Humboldt, and Goethe. The book concludes with an extended epilogue in which Richards argues that key elements of the Romantic view of life are retained in Darwin's own theorizing and, in consequence, that we misunderstand the core content of Darwin's theory when we try—as most contemporary theorists try—to conceptualize evolutionary theory without a prior commitment to an antecedent, form-giving archetype that underwrites the unity of forms we observe in large classes of organisms. He says, "In this epilogue I will try to demonstrate that the usual interpretation of Darwinian nature is quite mistaken, that Darwin's conception of nature derived, via various channels, in significant measure from the German Romantic movement, and that, consequently, his theory functioned not to suck values out of nature but to recover them for a de-theologized nature. I will indicate how his Romantic assumptions led him to portray nature as organic, as opposed to mechanistic, and to identify God with nature, or at least to reanimate nature with the soul of the recently departed deity" (Richards 2002, 516).

A couple of pages later he says, "When Darwin abandoned the Creator God, he did not, as I will argue, eviscerate living nature of teleological structure; rather his nature had exactly the same Romantic outlook as that depicted by thinkers like Alexander von Humboldt. And creative power would be transferred to nature in the gradual, evolutionary unfolding of telic purpose" (Richards 2002, 518). This is a remarkable thesis. If Richards is right, virtually all contemporary interpreters, myself included, are mistaken in attributing to Darwin a blindly mechanistic view of natural selection. But is Richards right? In particular, is it correct to interpret the perpetuation of Romantic metaphors as evidence that Darwin conceived of natural selection not in mechanistic terms but rather in terms of an unfolding, cosmic, telic purpose? Or is it perhaps more plausible to read the Romantic swells of Darwin's voice as evidence that he was indeed the shrewd rhetorician I have been trying to portray?

I cannot pretend to settle the issue here. Doing so would require close examination of far more passages from Darwin's mature works than those considered by Richards.[3] Nonetheless, the contrast between my view of Darwin as rhetorician and Richards's view of Darwin as Romantic is helpful in this respect. Richards is surely right to insist that any adequate interpretation of Darwin must explain the persistence of the Romantically charged language. It is simply implausible to think that the metaphors, cadences, and tone of the *Origin* were accidental. It is also true that, early in his intellectual development, and especially during and in the immediate aftermath of his five-year voyage on the *HMS Beagle,* Darwin fell under the spell of Humboldt's poetic descriptions of nature, based on Humboldt's own voyage a few years earlier. And, as Richards insists, it rings false to think that these powerful Romantic influences during Darwin's formative years were washed out of his intellectual orientation with no remaining traces. If, then, you are convinced (as I am) that Darwin's vision of evolution by natural selection is *not* of a piece with Romantic theology, then the burden rests on us to explain the obvious Romantic refrains in his voice. And this makes the view of Darwin as rhetorician look especially attractive.

My grounds for doubting Richards's main thesis are the following. During his years aboard the *HMS Beagle* and the years leading up to the publication of the *Origin,* Darwin's view was under construction. His theory of the diversity and adaptedness of living forms was slowly taking shape. It thus is hardly surprising, given his immersion in the works of Humboldt, Goethe, and others, to find him struggling to work out his ideas within the framework at his disposal. Indeed, Richards's case is strongest when discussing Darwin's letters, notebooks, and books from the 1830s all the way up to the publication of the *Origin* in 1859. By contrast, when Richards turns to passages from the *Origin* and especially passages from the *Descent of Man* (1871) and *The Expression of the Emotions in Man and Animals* (1872a), the case is weaker. The Romantic metaphors and cadences have not disappeared altogether, to be sure, but they have diminished in number and exuberance. More importantly, we do not find in Darwin's later writings the explicit endorsement, as we do in the earlier sources cited by

Richards, of the core theoretical substance of the views embraced by Humboldt and Goethe.

To the contrary, we find clear assertions that appear to entail the denial of the Romantic conception of life. Thus, for instance, where Goethe posits a form-giving archetype—an antecedently existing template for all living types—we find Darwin insisting on the formative powers of descent with modifications over great stretches of history: "On my theory, unity of type is *explained by* unity of descent" (Darwin 1859, 206; my italics). Darwin's explanation for the origins and perpetuation of discernible living forms appeals not to an antecedently existing form-giving source but simply to the effects of "slight differences accumulated over many successive generations." Descent with modifications is the wellspring of all the forms that have emerged and evolved. Those that exist today, like those that existed before, are combinations that have taken hold. They are accidental forms preserved for a while by natural selection. They are also vulnerable forms, the products of causal factors scattered across space and time, readily altered or extinguished by changes in any number of contingent states of affairs.

It seems to me, in fact, that at least some of Darwin's rhetorical strategies work precisely against Richards's thesis. This is evident in so far as Darwin's strategies are opposed to two failures of imagination within the Romantic view. There is, first of all, a failure of historical imagination like that of the

Hereford breeder. Part of Goethe's scientific genius was discerning morphological similarities across organisms in very large classes[4]—Goethe is, according to Richards, the inventor of the study of morphology—and these early discoveries in morphology were well-known to Darwin. But discerning such similarities across so many forms of life is one thing; explaining them is another matter. And as I have already pointed out, while Darwin's explanation is historical all the way down, Goethe's is not. Goethe certainly believed that his postulated archetype was perpetuated over generations, but, unlike Darwin, he did not give a historical explanation for the emergence of the archetype. To the contrary, for Goethe, the archetype had to exist antecedently in order to explain the evolution of life. By contrast, for Darwin, the mechanisms involved in the evolution of life explain the emergence of more or less distinct types. And this contrast vividly illustrates the extent to which Goethe and others underestimated the form-giving powers of accumulated changes over thousands or millions of generations. That is the first failure of imagination in the Romantic view.

The second failure runs in the opposite direction. The problem is not a failure to compensate for an impoverished historical imagination but rather the unchecked glorification of our aesthetic capacities, including our creative imagination. Consider for a moment the Romantic naturalist who believes that artistic insight is integral in discovering the (presumed) unity of all life. She is presented with the patterns of similarity discerned by Goethe. What does she do? Instead of cultivating the capacities of reason to compensate for an impoverished historical imagination—instead of following the route recommended by Darwin—she lets loose her powers of imagination. After all, although she readily discerns the patterns of similarity through observation, she cannot observe the correct explanation of those patterns. To borrow from Humboldt, the correct explanation belongs to a "distant zone" that we cannot travel. So we must compensate. We must "animate" that distant zone in our imagination so as to grasp the correct explanation with all the "vividness of truth." Yet, as Darwin insists, overly zealous imaginations seduce us away from the truth as surely as impotent imaginations. The Romantics are right that our knowledge of the truth ought to be viscerally alive—we should feel as well as believe—but vivacity alone must not become our criterion of truth. That, presumably, is why organs of extreme perfection appear an obvious counterexample to Darwin, but it is also why, in reply, Darwin urges that we check our imaginations by summing up in our minds the cumulative effects of small changes wrought over thousands or millions of generations.[5] As the power and reach of our historical imagination expand, the allure of the Romantic imagination, and with it the Romantic archetype, withers and dies.

I do not claim that these considerations are sufficient to settle the difference between Richards and me. Nor am I particularly concerned to resolve the issue in either direction. I am content to acknowledge the importance of Darwin's relationship to his Romantic predecessors and offer the hypothesis that the Romantic strains in his voice are best construed as evidence of his rhetorical insights. After all, although I do not agree with Richards, our views are nonetheless compatible. That is, if he were right that Darwin's metaphors reflect the substance of his view and not merely a rhetorical strategy, then, in addition to learning something terribly surprising about one of our greatest scientists, we would conclude that Darwin is indeed in the business of trying to help his readers see and feel his theory aright. We would not regard Darwin's use of these devices as a rhetorical ploy, of course, since bursts of effusive poetry are integral to a Romantic orientation. But sincerity does not make an effective rhetorical device any less rhetorical or effective. So even if the central tenets of Darwin's theory commit him to the metaphysics of the German Romantics, there is something of value to take from Darwin's mode of discourse. The aim of this book is to do just that.

Conclusion

There are, then, two deficits of the imagination that inhibit our attempts at understanding the theory of evolution by natural selection. Our overall orientation toward the world tends to be ahistorical; we tend not to notice the biasing effects of our cultural ancestry. That is the first deficit. And our lack of a historical sense reinforces and is reinforced by a range of biases to which we are naturally inclined, thanks to the architecture of our cognitive and affective capacities, including the tendency not to notice conflicts within our minds. That is the second deficit. Darwin was alive to these deficits throughout the whole of the *Origin* as illustrated in the passages above. It is precisely this Romantically inspired sensitivity to our habits of thought and dispositions of mind that helped make Darwin so shrewd a rhetorician without being a Romantic.

Darwin's rhetorical insights have not lost their value. They have, however, been insufficiently appreciated among philosophers and psychologists of scientific inquiry. My aim is to develop and extend his insights in ways that he would have appreciated, in light of what is currently known about our history and our neurological and psychological constitution. We can, if I am right, flesh out in some detail the failures of imagination to which Darwin draws our attention and, in so doing, begin the task of describing a naturalistic orientation toward inquiry that stands some chance of correcting the habits and dispositions that thwart our pursuit of knowledge.

The most vivid way to develop the power of Darwin's insights is to apply them to topics about which we are most likely led astray by our history and constitution. Perhaps there are several topics that fit this bill, but among the problems that have vexed our predecessors and continue to vex us the most is ourselves. Accordingly, my lead question, to which I now turn, concerns the nature of being human. It is this: What are we?

TWO

Our Most Vexing Problem: Conceptual Conservatism and Conceptual Imperialism

Man has always been his own most vexing problem. **REINHOLD NIEBUHR, *THE NATURE AND DESTINY OF MAN***

Niebuhr's answer to my lead question—What are we?—is that we are our own most vexing problem. And that is surely right. Nevertheless, Niebuhr's characterization of the nature of our vexation could hardly be further from the truth, and it is instructive to see where he goes wrong. The root cause of Niebuhr's error, I will argue, is that he suffers from the failures of imagination diagnosed by Darwin. Of course, Niebuhr is hardly unique in this regard. I have made him my first foil because he is an articulate and particularly learned representative of the types of failures that retard or otherwise minimize a naturalistic view of ourselves. Nor has Niebuhr's error been eradicated among serious inquirers of our day. As I argue in later chapters, his error is perpetuated in the works of some of our most visible theorists, many with expressed commitments to naturalism.

To make my case against Niebuhr, I address issues at two levels of inquiry. One level concerns the substance of human agency and addresses my lead question directly; the other concerns methods of inquiry best suited for discovering the nature of human agency. I proceed this way because I see no plausible alternative. The best methods for discovering the truth about ourselves are methods framed, at least in part, by Darwin's rhetorical insights. Our methods of inquiry ought to be informed by what we already know or rationally believe about ourselves, especially the limitations and illusions we bring to the task. We should feel impatience toward methods uninformed by such knowledge. At the same time, our best theories about ourselves are only as fruitful as the methods with which they were generated and sustained. Theories of human nature, including theories in neuroscience and psychology, must be informed and constrained by what we already know of the fecundity of various methods or strategies of inquiry. It is hard to see how to make progress on questions of substance without simultaneously addressing questions of method. About that, Nietzsche was surely right.

It goes without saying that there may be better approaches to the study of human agency. The shortcomings in my approach may be too much. Still, every attempt at serious inquiry begins with a bet: which horses are likely to cross the finish line at all and which one is likely to cross first? In this regard, it is crucial to observe that our present-day understanding of human agency is at a critical crossroads. Though we do not know with high confidence what kind of agents we are, we do know with overwhelming confidence what kind of agents we are not. We know in light of our history and in light of discoveries concerning our constitution that we are not the kind of agents that our Enlightenment and Romantic predecessors took us to be. We also know that the way forward cannot be a simple matter of accumulating more information, for our methods are in need of reform. Today we are learning at a rapid pace how the effects of our history and our constitution bias or blind our view of ourselves, and some of the relevant discoveries are canvassed through-out my discussion. It is in light of these discoveries that the prospects for further knowledge of human nature appear far greater, in part by developing methodological strategies for diminishing the tendencies in ourselves that bias and blind us. And that, once again, is why we cannot help but focus our efforts on the messy back-and-forth between method and substance.

Niebuhr's Antinomies

To focus discussion, I begin with a few thoughts from one of the great theologians and cultural historians of the twentieth century. Although it may appear disingenuous or hyperbolic to begin a book on naturalism and agency with an excursion into theology, I intend to convince you otherwise. For among the various orientations toward inquiry currently in fashion, including those that boast of naturalistic credentials, nearly all are informed by more than just a little theology. Once this becomes clear, we will be well positioned to see how to extend and apply Darwin's rhetorical insights.

Niebuhr begins his magnum opus, *The Nature and Destiny of Man,* with the declaration that serves as epigraph to this chapter. He continues, "How shall he [the human agent] think of himself? Every affirmation which he may make about his stature, virtue, or place in the cosmos becomes involved in contradictions when fully analyzed. The analysis reveals some presupposition or implication which seems to deny what the proposition intended to affirm" (Niebuhr 1942, 1). Human nature, according to Niebuhr, is conflicted. The conflict is that although we belong to the natural order—the human being, he says, is "a child of nature, subject to its vicissitudes, compelled by its necessities, driven by its impulses" (Niebuhr 1942, 3)—we are also above or beyond or outside the natural. Naturalists are right to insist that we are part of the animal kingdom, but they err in thinking that the facts of our animal nature suffice to explain the full range of human capacities:

...man is a spirit who stands outside of nature, life, himself, his reason and the world. This latter fact is appreciated in one or the other of its aspects by various philosophies. But it is not frequently appreciated in its total import. That man stands outside of nature in some sense is admitted even by naturalists who are intent upon keeping him as close to nature as possible. They must at least admit that he is *homo farber,* a tool-making animal. That man stands outside the world is admitted by rationalists who, with Aristotle, define man as a rational animal and interpret reason as the capacity for making general concepts. But the rationalists do not always understand that man's capacity involves a further ability to stand outside himself, a capacity for self-transcendence, the ability to make himself his own object, a quality of spirit which is usually not fully comprehended or connoted in "ratio" or "nous" or "reason" or any other concepts which philosophers usually use to describe the uniqueness of man. (Niebuhr 1942, 3–4)

No philosophical theory, not even the most extravagant rationalism, does justice to the human capacity for self-transcendence, a capacity beyond mere reason. What has vexed our predecessors and what vexes us is the apparent impossibility of integrating this remarkable capacity with our status as "children of nature." Every attempt at integration has fallen prey to internal conflicts, to what Niebuhr calls irresolvable antinomies, and precisely this is the source of our most vexing problem. Or so he says.

But we should wonder. Niebuhr asserts that we are endowed with the capacity to (1) take ourselves as objects of reflection or inquiry, (2) stand outside ourselves, (3) transcend ourselves, and (4) exercise a quality of spirit that is distinct in some way from reason. And he says this with confidence, as if it is all so obvious and clear. The only obvious point, however, is that Niebuhr's claims are far from clear. We are left to wonder, in the first place, what on earth he might be talking about, since none of the terms in 1–4 is explicated. It is also hard to know what evidence he might offer that should move the naturalist or even the nonnaturalist. After all, everyone can grant that humans are tool makers and tool users without granting that we are outside nature, since being unique in nature hardly entails being outside it. (There is, of course, the further point that it is empirically false that humans are the only tool-making or tool-using species on earth.) The naturalist can also agree that we have the capacity to take ourselves as objects of thought, since there is nothing nonnatural or supernatural in conceptualizing ourselves or some part of ourselves in the same way we conceptualize other objects of thought. But perhaps the most puzzling aspect of Niebuhr's claim is simply the overweening confidence with which he tosses off the assertion that we are self-transcending "spirits." Whence his confidence that it is part of our nature to transcend the natural by ascending to something above nature?

This is not a trifling question, since the answer takes us to the heart of Niebuhr's overall orientation toward inquiry. The confidence with which he describes the antinomies in our nature stems not from his instincts as a critic or historian of culture but rather from his orientation as a theologian. To see that this is so, consider the naive question, What is the goal of theological reflection? I am not asking about the goals of religious rituals such as confession or prayer or communion, or the psychological and social functions of religious attitudes or states of mind such as reverence or guilt or contrition. I am asking instead about discourses by highly learned minds set forth in lectures and books, discourses that assert and appear to defend claims concerning the nature and even the destiny of human beings. The theological tradition to which Niebuhr belongs is in the business of seeking and disseminating grand truths about human nature. So the question to ask is, If Niebuhr's account of our most vexing problem belongs to a tradition of trying to discern the truth, why does his most crucial assertion concerning our nature stand so naked on the page, so unadorned by evidence?

The answer rests precisely on the general orientation at the heart of Niebuhr's methods of inquiry and, indeed, at the heart of all theological inquiry. It rests on the default assumption that, with respect to the most central features of being human, we begin the process of inquiry *already knowing what we are like.* The default is that we come to the activity of inquiry already knowing that some of the properties in terms of which we appear to conceptualize ourselves—'self-transcendence,' 'free will,' 'responsibility,' and so on—are constitutive of the kinds of beings we are. We know not only *that* these concepts are constitutive but also *what* these concepts dictate, that is, what our constitution must be like for these concepts to apply. The assumption is that we already know, for instance, that self-transcendence is indeed a defining property of being human and, moreover, that this property places us outside the natural order. According to this orientation, any theory that alters or fails to account for the concepts that we know to be constitutive of human nature is simply not a theory of us. It is, at best, a theory of something other than human nature. A commitment to this general orientation—that we already know, in advance of inquiry and at least in outline, what a successful theory of human nature must look like—explains the confidence with which Niebuhr insists that we are a conflicted mix of nature and something other than nature.

The default assumption at the heart of theological reflection has, moreover, a direct analog in most contemporary philosophical reflection. The analogy can best be seen by considering the basic aims of philosophical reflection. As most of us practice it today, the goal of our reflections is to identify concepts of apparent importance, especially concepts such as 'freedom,' 'responsibility,' 'goodness,' 'consciousness,' and so on that bear on the nature of being human and then to locate these concepts amid the concepts and claims of our best developed sciences. This is to engage in what I will call the *concept location project.* The aim is to integrate our humanistic self-understanding with our emerging and evolving scientific view of the world. This, no doubt, has been one of the most pressing intellectual tasks since the rise of modern science, but in pursuing this project today, most philosophers, like the theologians, tend to endorse the default assumption described above. When we are the topic of our reflections—when the task is to understand the nature of being human—we tend to assume that we already know, in advance of scientific inquiry and at least in outline, what we are like. We tend to assume that the concepts of freedom, responsibility, goodness, and so on that we bring to the table are constitutive of what we are and that we already know or can readily discern what we must be like in order for these concepts to apply to us.

The central target of my discussion then is the default assumption of all theological and most philosophical reflection, and, as I have indicated, we will discover that some of our most naturalistically inclined theorists nevertheless fall back upon this assumption at crucial points in their reflections. The constructive aim of my discussion is to suggest strategies for finally eradicating the residual theology that retards the progress of human knowledge.

Of course, most theorists who are overtly *non* naturalistic nonetheless tip their hats to the astonishing growth in human knowledge since the rise of modern science. But they do so in a defensive posture, and their strategy always rests upon the same ploy. In one form or another they strive to minimize the unmatched power of scientific methods by subordinating those methods to the default assumption just described. The scientific pursuit of knowledge is constrained by or beholden to the authority of concepts that appear to play an especially important role within our worldview. This orientation, as I say, is at the heart of all theological and a good deal of philosophical discourse. To help fix our ideas, then, I am going to describe two general orientations toward inquiry within the contemporary intellectual landscape (I describe a third in a later section). Both rest upon the presumption that, with respect to important features of human nature, we already know what we are like. I then return to Niebuhr's view of self-transcendence, for we may, armed with both orientations, appreciate how starkly the theologian's orientation, as well as the philosopher's concept location project, contrasts with a robust naturalistic orientation.

Two Orientations Toward Inquiry

The first orientation, what I call *conceptual conservatism,* comes in two distinct species. These include

1. the commitment to preserve as far as possible any concept that appears important within some well-developed scientific theory
2. the commitment to preserve as far as possible any concept that appears important within our general worldview

Niebuhr illustrates the commitment to the second orientation with respect to 'self-transcendence' and, more generally, 'human agency' as these concepts function in our cultural worldview.[1] Indeed, Niebuhr belongs to a tradition in which the long genealogy of any concept in our worldview is taken as powerful evidence of its importance. The longer the historical roots, the more likely a given concept designates something of substance—or so it is assumed; and the greater the appearance of substance, the more pressing that the concept be "located" and thereby preserved. Concepts with the longest roots are those that must be retained by reconciling them with emerging scientific knowledge. The perpetual task of theology, given this orientation, is to preserve our most venerable concepts by continually integrating and reintegrating them with everything else we believe to be true of the world. And concepts concerning the nature of human agency are, as a matter of historical fact, among the most venerated.

The fixation in Niebuhr (and, as we will see, in several self-avowed naturalists) on the long, persistent history of a concept may appear a sensible way to proceed. Longevity, it may be thought, is evidence that the relevant concept is tracking something important and real. The problem with conceptual conservatism, however, is not a simple-minded distrust of concepts with long ancestral roots. The problem is more specific. Conservatism is objectionable with respect to what I call *dubious concepts.* One category of dubious concepts comprises concepts that descend to us from a worldview we no longer regard as true. A concept is dubious by descent if, for example, its genealogy traces back to the theological worldview of our Romantic and Enlightenment ancestors—a view we no longer regard as true or promising—and if the concept has not been vindicated in any well-confirmed scientific theory. This is not to assume that all theological claims are false, only that none has shown itself relevant to the enormous progress in knowledge since the rise of modern science. The steady, increasing irrelevance of theological concepts and claims to progress in inquiry is, I take it, as near a brute fact as we are likely to get in the study of history. It thus is rational to frame our intellectual problems and solutions in terms that leave out or neutralize concepts dubious by descent. Precisely this, I suggest, is behind Darwin's concern to identify and minimize the effects of certain culturally inherited habits of thought.

There is, moreover, a second category of dubious concepts, namely, concepts dubious by psychological role. Some of our most basic conceptualizing capacities lead us astray. Our tendency to overapply the perceptual concept 'face'—to clouds, automobile grilles, abstract paintings, and so on—is a vivid illustration. But there are other, more substantive cases. Several contemporary psychologists hypothesize that the human mind comprises at least two ensembles of conceptualizing capacities, one concerned with causal relations, the other with mental states. Our cognitive and affective capacities, it is claimed, are structurally disposed to apply the concept 'cause' under a range of perceptual conditions, several of which involve nothing more than the mistaken appearance of a causal relation. We are also structurally disposed to apply a host of mentalistic concepts, including 'intention' and 'belief,' even when the attributed mental state does not exist. Or so it is claimed.[2] And concepts for which our conceptualizing capacities are like this—apt to generate false positives—are dubious by psychological role. This, I suggest, is behind Darwin's concern to identify and minimize the effects of certain inherited dispositions of mind. The problem with conceptual conservatism then, as we will see throughout this book, arises with respect to concepts dubious by descent, by psychological role, or by both.

The second orientation toward inquiry is what I call *conceptual imperialism,* namely, a commitment to the view that certain concepts have a right among and dominion over all other concepts and all methods of inquiry. This is not a commitment to saving as far as possible the relevant concepts, since the question of altering or eliminating certain concepts has been closed off from all discussion. Instead, the task is to reflect upon the content of such concepts and, on the basis of such reflections, describe what the world must be like for it to fall under such concepts. The assumption is that we already know, at least with respect to certain concepts, everything we need to know for their correct application; the only question is whether the actual world fits the properties specified by the content of the concepts.

Lest anyone suspect me of caricature in the case of imperialism, consider the imperialist's orientation toward the concept 'free will.' The imperialist presumes that some apparently important concepts are necessary in some sense and hence nonnegotiable, at least at their core. The claim concerning 'free will' is that if we alter or relinquish our traditional concept we thereby lose our grip on a host of concepts associated with the notions of control and responsibility. Altering or eliminating those concepts eviscerates our conceptual scheme to such an extent that we can no longer think or offer meaningful assertions about human agency. We will have changed the subject and lost our way. From the imperialist's point of view, there is something inherently self-destructive about a form of inquiry in which a concept so near the core can be altered or eliminated. What, then, is the positive aim of imperialist inquiry concerning free will? The aim is to reflect on our concept by way of thought experiments and articulate conditions necessary and sufficient for its application. Once we have accomplished that, we may then specify what the world must be like for something actually to fall under the concept. On the imperialist assumption that our concept must be retained, and on the basis of conditions deemed necessary and sufficient, we are forced to conclude, as Roderick Chisholm (1964) does for example, that human beings must be akin to God with respect to the will. The imperialist claims in this case to have discovered that the core of the human self must include the capacity to initiate causal sequences leading to free actions without that capacity itself being subject to efficient causation. This is a claim about the nature of human beings discoverable prior to actual inquiry into the workings of the human mind, a claim based solely upon the alleged nonnegotiable content of our alleged concept 'free will.'

The imperialist orientation is, in this regard, more audacious than that of the conservative. While the conservative is committed to saving certain concepts as far as possible consistent with other theoretical and empirical discoveries—where "consistency" is open to more or less stringent interpretations—the imperialist is committed to saving certain concepts come what may, no matter the violence with which discoveries in other areas of inquiry are distorted or simply ignored. Of course, there are degrees of gradation between the two. It is possible to be more or less conservative with respect to some concepts while being more or less imperialistic with respect to others.

It is also important to notice that imperialists are typically most imperialistic toward dubious concepts. Indeed, it is precisely because the concepts are dubious—because of their long, persistent history or because of their recurring salience in our psychology—that the imperialist feels so confident that the concepts are not open to negotiation. It is understandable why so many theorists proceed in this manner. If we fail to reflect on Darwin's insights—if instead we take it for granted that the way we are naturally inclined to conceptualize the world reflects the way the world really is—then concepts with long historical roots and concepts that run deep in the fabric of our psychology are going to rise to the top every time. It will feel utterly natural, utterly correct, to afford such concepts substantial authority in the way we frame our intellectual tasks. It is only when we take the naturalistic turn regarding ourselves that things begin to look different. Once we see that many of the concepts we inherit from our predecessors, especially concepts concerning human agency, have their home in worldviews we no longer regard as true, it becomes obvious that the authority those concepts once wielded must now be withheld. And once we learn the myriad ways we are led astray by our own affective and cognitive capacities, especially by capacities that exercise various concepts concerning human agency, it becomes similarly obvious that the same divesture of authority applies to concepts dubious by psychological role.

Niebuhr appears a robust imperialist concerning 'human agency' in much the way that Chisholm is regarding 'free will.' The charge that Niebuhr levels against naturalists and rationalists is that they persist in ignoring what appears to be an essential feature of being human, namely, the "spiritual" capacity for self-transcendence. And it is precisely the imperialist strains in his orientation that provide Niebuhr the assurances he needs that no argument or evidence is required in support of this charge. His default assumptions include the following: we can know in advance of scientific inquiry that human nature essentially includes a nonnatural capacity to overcome our merely natural capacities; we can know in advance of scientific investigation that this essential capacity is not fully captured within the bounds of human reason; and so, in consequence, we know that any theory of human agency that fails to explicate and thereby preserve this integral capacity is an unqualified failure. The imperialist presumption in all this is that nothing we discover about the world can alter the requirement that, in order that anything qualify as a human agent, it must be endowed with the capacity for spiritual self-transcendence.

Our Most Vexing Orientation

On Niebuhr's view then we are manifest contradictions. Human nature comprises but cannot combine the spiritual and the natural, and precisely this vexes us most. But—need it be said?—Niebuhr is wrong. He is wrong to think we are self-transcendent spirits and wrong to insist we suffer antinomies between what is natural and what is not. And in a lovely bit of irony, it is precisely Niebuhr's failure, his misdiagnosis of our most vexing problem, that reveals the actual causes of our most vexing problem. What causes us to vex ourselves most is not the persistent failure to integrate the spiritual with the natural but rather the persistence of that very way of seeing ourselves! We vex ourselves because we persist in framing our inquiries in terms of concepts concerning the human self that are patently dubious. Our most vexing problem is caused not by Niebuhr's antinomies but rather by the orientation toward inquiry that leads him to his antinomies. It is, in short, the theologian's orientation—conservatism and imperialism regarding dubious concepts—that vexes us most.

Niebuhr's mistake, and the mistake of all theologians, is the antecedent commitment to saving dubious concepts concerning human agency. Such a commitment is a strategic error, for it binds us to a theological worldview we no longer regard as true and invokes a host of psychological tendencies that tempt us away from the truth. This is the rhetorical insight to which Darwin appeals repeatedly throughout the *Origin;* this is the insight he hoped would ease the reception of his theory. After all, no one who is genuinely concerned to discover the truth about the world would knowingly frame an intellectual task in such a way that our historical and constitutional infirmities are left in their usual places. Genuine inquiry requires that we resist certain parts of our nature as well as certain effects of our history. That Niebuhr's theological orientation embodies this sort of strategic error is a claim I defend as I proceed. Here, however, I pause to give some feel for the sorts of considerations offered in support of this charge.

Begin with a single thread in the history of the concept 'self-transcendence.' We have the capacity for self-transcendence, or so it is said, but what does this come to? The claim, I take it, is that we are capable of adopting a perspective on ourselves that affords us an extraordinary degree of epistemic strength and practical control. We ascend to such levels of strength and control, presumably, by rising above the limits imposed by the contingencies of our constitution. Self-transcendence thus construed is a radical form of freedom from constraints typically associated with our status as animals—the "vicissitudes," "necessities," and "impulses" of nature—all those nasty things that might wrest from us our presumed self-control. The further we ascend out of our animal nature, the better we achieve a godlike perspective, a point of view utterly unconstrained by contingent matters. Such a view, needless to say, runs deep in our theological traditions; we are "spiritual" the closer our kinship to god. This is how Descartes (1641), for example, describes human freedom. He claims that our will is as free and infinite as God's; he claims that it is precisely with respect to our infinite will—an unconstrained capacity to act or refrain from acting in the face of inclination—that we most clearly bear within ourselves the image of God. Although nothing short of earthly death can free us entirely from our animal constraints, we can, according to Descartes, reliably discover contingent and eternal truths by prudent exercise of our free will. We can, that is, by virtue of godlike self-control, aspire to knowledge that is similarly godlike.

Even this tiny fragment of the genealogy of the concept 'self-transcendence' ought to give us pause. Descartes locates our capacity for self-transcendence in freedom of the will and then locates our capacity for free will in our kinship to the Judeo-Christian god. Self-transcendence, for Descartes, is a capacity we have in so far as we were created in the image of God. But today we can no longer accept that the human will is akin to the will of God; discoveries in psychology and neuroscience, and the fact that we are evolved animals, all tell against it. We certainly cannot accept the image-of-God thesis prior to actual inquiry, and although the thesis that we, in some capacity or other, bear the image of God may be logically compatible with the facts of our constitution and history, such compatibility is not the point. The scientific and historical evidence points overwhelmingly to a godless explanation of the origin and nature of the human capacity for practical reason. And it is rational to accept explanations supported by the preponderance of available evidence.

It is perhaps worth harping on the point about history a moment longer. Darwin in the *Origin* argues incessantly against the creationists of his day. He repeatedly faces the question, "Why will my peers and people in general resist the theory of evolution by natural selection?" The question is a rhetorical one, in the sense that the correct answer suggests a strategy for maximizing the chances of convincing one's opponents. And as we have seen, Darwin gives the simple but compelling answer that most of us suffer from an undeveloped historical imagination: we fail to sum up in our minds the accumulated effects of small changes over thousands or millions of generations. That is the failure of the Hereford cattle breeder.

The thing to notice is that the same point that Darwin wields against the Hereford breeder also applies to the effects on us of concepts dubious by descent. We bring to inquiry a host of concepts bequeathed to us by our intellectual predecessors and tend to expect that the most important among them will surely be preserved. That is simply how we conceive our intellectual task—the preservation of apparently important concepts. Like the Hereford breeder who fails to imagine the countless historical changes by which his preferred animal evolved, we fail to appreciate that all our concepts, including those we take to be descriptive of what we essentially are, are the evolved products of our animal history. We also fail to appreciate the powerful inertia of culturally instituted conceptual categories and the features of our psychology that anchor those cultural institutions. We fail, that is, to sum up in our minds the accumulated effects on our current conceptual scheme of changes over thousands of generations, and we fail to gauge the resistance to change that our culturally entrenched categories possess. And this makes it virtually inevitable that we will underestimate the vulnerability of our concepts to the "vicissitudes" of future changes.

Turn now to what is known of our affective and cognitive dispositions and the bearing such knowledge has on our attitudes toward our concepts of human agency. The capacity for self-transcendence, as noted, is presumably the ability to achieve an extraordinary degree of epistemic insight and practical control. Yet the facts of the matter tell heavily against this happy view. This is clear from considering but a single point upon which contemporary cognitive psychologists and neuroscientists insist, namely, the extent to which our mental lives are lived mostly beneath the level of conscious awareness and beyond the reach of conscious volitional control. As I describe in later chapters, the hypothesis that most of our psychological capacities operate below the level of awareness is increasingly compelling from empirical and theoretical work on perception, memory, emotions, practical reasoning, and more, and this hypothesis casts doubt on our alleged capacity for self-transcendence. For some psychological capacities, the challenge is that reliable first-person access is difficult or too costly; for others, the problem is that such access is not possible. In general, the only way to acquire reliable knowledge of one's mental life is to achieve a perspective external to one's first-person point of view by first developing a scientific orientation toward knowledge of one's self.

There is, moreover, an additional point against the claim concerning self-transcendence, namely, that we are subject to a variety of constitutional conflicts. These conflicts arise from the fact that, while some of the systems that constitute our minds are responsive to consciously accessible inputs, many are not responsive in that way, and often the two sorts of systems operate independently of one another. We have, for example, a system that responds to conscious thoughts about actions we wish to perform and to conscious perceptions of ourselves performing certain actions. This system does not operate, so far as we can tell, in response to the low-level, nonconscious processes that actually cause us to act. The evidence indicates that the first system, the one responsive to limited conscious inputs, sometimes (and perhaps most of the time) operates independently of the other, low-level systems that generate our actions. That, at any rate, is the provocative thesis defended in Daniel Wegner's marvelous book *The Illusion of Conscious Will.*[3] And to the extent that Wegner's theory and others like his are plausible, we must take seriously the possibility that even when we appear to be acting out of careful deliberation, the real causes of our action do not correspond to the reasons we give ourselves or others for our action. That is, we must take seriously the possibility that our sense of control and self-knowledge, even in ordinary cases, is an illusion engendered by a conflict between distinct systems within our psychology. Seen from this vantage, a vantage afforded by our best contemporary theories, the claim that we are capable of transcendent self-knowledge or control is extraordinary. It is a fanciful view of ourselves, the sources of which need to be divested of the authority they wield in the formulation of our intellectual tasks.

One way to appreciate the force of Darwin's point concerning our cognitive and affective dispositions is to consider the following question. If the cognitive and affective systems responsive to consciously accessible thoughts or feelings operate independently of other cognitive or affective systems (including those that cause us to act), what is it rational to expect regarding our knowledge of ourselves? And one answer to this question is that we should not expect of ourselves that we can reliably see or otherwise access the actual causes of our actions from the first-person point of view. We should expect, to the contrary, that the way we appear to ourselves—the contents of our consciously accessible thoughts and feelings—is partial or misleading and in some instances altogether mistaken. We should expect that, in order to know ourselves, we must approach our selves using our best scientific methods and, in particular, resist the default assumption that we already know on the basis of first-person considerations what we are like.

And this expectation underscores a failure of imagination parallel to the one described above. Darwin's cattle breeders saw the world with undeveloped historical eyes, but it is at least as likely that we are similarly blind to relevant facts of our constitution. Most of us live with considerable confidence that we have, or that we can achieve, substantial introspective knowledge of why we do what we do. Most of us think we acquire such knowledge in ways that are obvious and unproblematic—by reflecting inwardly, for example, or by simply pausing and taking note of what is going on. But most of us are wrong on this score.[4] Thus, in addition to an undeveloped historical imagination, we also suffer from an undeveloped psychological imagination, namely, the inability to see past our illusions of self-knowledge to the depths of our ignorance regarding our own agency.

So Niebuhr is right that we vex ourselves, but he is wrong that we do so by way of his alleged antinomies. It is, instead, Niebuhr's own orientation toward inquiry—his commitment to conceptual conservatism and imperialism, the mark of all theology—that is the real root of what vexes us most. It is the persistence of conservatism and imperialism with respect to dubious concepts that stifles genuine inquiry. And it is the explication and defense of an alternative orientation that offers the hope for a form of inquiry that leads to progress in knowledge.

A Third Orientation

The third orientation toward inquiry is what I call—with no pretense of impartiality—*conceptual progressivism.* This orientation is at the heart of the naturalistic methods of inquiry I wish to advocate. According to the progressive orientation, inquiry is largely a matter of exploration. Inquiry is exploration in the way that the naturalists of the nineteenth century were explorers—Alexander von Humboldt, Alfred Wallace, and Charles Darwin, for example. To adopt this orientation is to embark on the study of natural systems with the positive expectation that our settled view of the world is about to be unsettled. It is to proceed with a heightened sense that, as we approach the bounds of current knowledge, we are crossing into lands that may challenge the very categories in terms of which our knowledge is presently expressed. This is especially so for knowledge couched in concepts dubious by descent or by psychological role. Indeed, to adopt a progressive orientation toward dubious concepts is to face the crucial possibility that we are conceptually ill equipped to assimilate new discoveries, that we may be thrown into conceptual disarray and forced to flounder, even forced to create new conceptual categories to make our way forward. Genuine inquiry for the progressivist is always a looming threat to our current worldview and a looming challenge to our creative powers.

Progressivism is the antithesis to conservatism and imperialism. The point of progressivism is not to preserve what appears important or unalterable but to push back the bounds of our ignorance by conforming our concepts to the world as far as possible. And, in light of what we know about our constitution and our history, we should expect to live through periods of conceptual unease and confusion. We should expect that what feels or appears important to us is often not important; our own capacities sometimes lead us astray. We should also expect, given what we know about our history, that our intuitions and hunches are calibrated in ways that often lead us astray. These sources of knowledge concerning our constitution and our history underwrite the expectation that many of our most important concepts—those concerning human agency, in particular—do not conform to the way the world in fact is. The aim of progressivism is to change this fact. The aim is to conform our selves—our concepts, our thoughts, our feelings—to better fit the nature of realty.

When the progressive enters the lab or the field, the task is not to call everything into doubt. I do not suppose that the device of systematically doubting all of our beliefs is a coherent possibility. I do not even suppose that it is possible to systematically doubt a large proper subset of all we believe. Nor is there anything about the progressive orientation that suggests anything so unrestrained. The progressive is alive first and foremost to dubious concepts—to concepts dubious by descent or by psychological role—and that means that the progressive, in order to correctly identify concepts that are dubious, must be informed about the genealogy of our concepts and the most compelling theories in neuroscience and psychology. Progressivism cannot be had for cheap; it is anchored by a grasp of what we currently know concerning our history and our constitution.

Nor is progressivism to be conflated with what we might call intellectual humility. Humility is certainly in order regarding the historical and constitutional infirmities that afflict us. But any further pretense of humility has no place in progressivism, for at the heart of the progressive orientation is a growing body of *knowledge* concerning the habits of thought and dispositions of mind that retard human inquiry. The progressive expects substantial change to our current conceptual scheme, without doubt; she expects it, however, not out of affected piety but as a consequence of all that is presently known about the kind of animal we are.

Although it is difficult to overstate the value of progressivism for inquiry into human agency, it is also easy to miss one of its central features, and this may lead us to overstate its value. The central feature is that progressivism as an orientation toward inquiry is entirely a product of its time and place. As long as our beliefs about our constitution and history upon which progressivism is predicated withstand the results of ongoing inquiries, the value of progressivism holds. But if the substance of what we know about our capacities or about our cultural and scientific histories should change, the elements of progressivism would have to be reevaluated. This is to say that no set of methods for human inquiry can be assessed outside the particulars of what we are like and where we have been. In so far as the details of our constitution and history rest upon a host of earthly contingencies, the methods most likely to lead us toward the truth must rest upon the same. How, given the kind of animal we are, could it be otherwise?

Whether or not we care to face it, we are subjects of the natural world. It thus is prudent to adopt the task of subjecting our thoughts and feelings as far as possible to the natural world. That we can know what we are like prior to actual inquiry is a particularly pernicious lie that we, thanks to the very structure of our psychology, are geared to tell ourselves. It is also a lie we are geared to believe. The task, therefore, of adopting the progressive orientation may contribute substantially to the growth of human knowledge, including knowledge of ourselves. It is, in any event, the horse on which I am placing my bets.

THREE

Naturalism as Exploration: The Elements of Reform

It is obvious that the concepts we choose to guide our experimental inquiries must be as flexible and profound as the functional processes that actually exist in nature. **JAAK PANKSEPP, *AFFECTIVE NEUROSCIENCE***

The orientation of all theology and a good deal of contemporary philosophy is deeply conservative, if not imperialistic, regarding traditional concepts of human nature. This orientation retards the growth of knowledge concerning our selves. The question to ask then is, Might we correct for the historical and constitutional facts that drive us toward conservatism and imperialism? The answer is that, at present, we simply do not know the extent of our powers to correct for the biases of our history or our constitution. Still, I shall wager that at least some degree of correction is possible, provided we adopt the quite different orientation of a progressive. I shall further bet that the best way to cultivate the progressive orientation is to explicitly adopt a set of directives in the way we frame and pursue our intellectual tasks. The main business of this chapter is to introduce the core directives I have mind, the power of which I illustrate by example in subsequent chapters. I conclude this chapter with an explanation of the relationship between my discussion of agency in theories of biological functions in part 2 and my discussion in part 3 of current theories of human agency.

Elements of Naturalism

Naturalism as exploration is not a metaphysical thesis. It is concerned rather to answer the question, By what methods are we most likely to discover the truth? Applied to the study of human agency, the question concerns methods most likely to produce knowledge of our nature. So far as I can see, and for reasons suggested in earlier chapters, among our most vital sources for answering this question are lessons culled from human history, particularly the history of modern science and the history of culture, and lessons concerning our constitution, particularly our neurology and psychology. Considerations drawn from our history and our constitution make plausible several directives for naturalistic exploration. I will introduce five of the more central directives here. Additional directives will be introduced in later chapters.[1]

Before I describe my directives, however, a brief word about tactics. The directives I offer are utterly banal, and intentionally so. I want constraints on inquiry based on considerations so uncontroversial that even the imperialist cannot reject them out of hand. That has an obvious strategic payoff when dealing with nonnaturalists. The downside, of course, is that serious naturalists may find me tiresome. But I beg your indulgence. For if I am right, the directives I offer, despite their banality, are not without power. There may even be room for serious naturalists to become better naturalists.

One more prefatory comment. I suggested in chapter 1 that among Darwin's rhetorical insights is an insistence on the retarding effects of certain habits of thought and certain dispositions of our minds. The aim of my directives is to put a bit of flesh on these admittedly bare bones. I do not claim that these directives are formulated anywhere in the *Origin.* Indeed, what motivates and substantiates some of these directives are discoveries made long after Darwin's death. I claim only that, first, they are plausible on their face in light of what is known today and, second, they are in the spirit of things that Darwin does say.

History of Science

The first directive derives from the history of science. We make progress in our knowledge of natural systems to the extent we *analyze inward* and identify low-level systemic mechanisms and interactions that instantiate high-level capacities. We also make progress as we *synthesize laterally* across related domains of inquiry, as we look for coherence among taxonomies of mechanisms postulated in associated areas of study. Such analyses and syntheses serve as checks on our theories and as mutual checks on one another. The importance of these checks is demonstrated throughout the history of science, including, for example, the evolution in our concept 'natural purpose' from Blumenbach (1781) and Kant (1790) to the modern synthesis in biology. The important claim is that the initial conceptual categories with which we begin such inquiry are often altered and sometimes destroyed as we analyze inward and synthesize laterally. It is a historical fact that as biology matured in the first half of the twentieth century, Bergson's (1907) 'life force,' a notion inherited from Blumenbach, Kant, and Kant's Romantic successors, could no longer be sustained because scientists discovered low-level physical mechanisms implementing precisely those capacities that, according to Blumenbach, Kant, and Bergson, could not be understood mechanistically.

A full defense of this claim requires detailed, historical case studies and, on the basis of such studies, it is reasonable to accept the following directive:

The expectation of conceptual change (EC): For systems we understand poorly or not at all, expect that, as inquiry progresses—as we analyze inward and synthesize laterally—the concepts in terms of which we conceptualize high-level systemic capacities will be altered or eliminated.

It is important that (EC) expresses an expectation and nothing stronger. But it is also important that the expectation expressed can have substantive consequences for the way we frame our intellectual tasks. We should expect our initial conceptual categories to change as we analyze into complex natural systems of which we are presently ignorant. And, in light of this expectation, we should calibrate our intuitions and hunches so that an explicit openness to change in our concepts becomes our default position. This is one way we may better subject ourselves to the world.

The power of (EC) and the lessons that motivate this directive are hard to exaggerate. Once we are primed to see them, we meet with illuminating instances at nearly every turn. We find them in abundance when we turn to the history of modern science. In *Discovering Complexity,* for example, William Bechtel and Robert Richardson (1993) describe the task of analyzing inward as a two-step process involving decomposition and localization. The bulk of their discussion comprises a wealth of detailed case studies illustrating the power of these steps, ranging from discoveries in the study of respiration and fermentation to studies of spatial memory, the nervous system, developmental genetics, and more. And although Bechtel and Richardson point to limitations of decomposition and localization when applied to certain types of systemic processes—parallel distributed processing, for example—it nevertheless appears plausible that we can analyze into such systems, including those with subsystems that run in parallel on spatially distributed computations, with appropriate expansions of the basic notions involved (for example, the concept 'mechanism').[2]

We need not, however, restrict our view to the sophisticated case studies offered by Bechtel and Richardson (and others) to appreciate the power of the lessons that motivate (EC). We need only read the newspaper! Science writer Carl Zimmer, for example, describes recent discoveries concerning the genome of the jellyfish that appear to undermine our previous theories of their evolution, their development, even their anatomy. The discoveries involved are really quite startling. The body of a jellyfish appears incredibly simple, comprising a relatively shapeless and undifferentiated web of nerves. Its basic body plan, as Zimmer puts it, has "the symmetry of a bicycle wheel, radiating from a central axis," which is to say that the jellyfish, unlike flies or fish or human beings, has neither a front nor a back, neither a left side nor a right side (Zimmer 2005). The body of the jellyfish thus appears a rather primitive thing. This appearance of primitive simplicity understandably led systematists to hypothesize that jellyfish were among the earliest animals to evolve, descending from simple, sponge-like organisms. On this evolutionary hypothesis, bilateral organisms, which are clearly more complex, emerged many millions of years after the jellyfish.

This hypothesis, however, appears to be badly mistaken. What is particularly striking is that the relevant mistakes have been uncovered as scientists have analyzed into the genetic structure of the jellyfish and then sought to synthesize their findings in genetics with theories concerning the embryology, anatomy, and evolutionary history of the jellyfish. The first discovery, based on the molecular clock of jellyfish genes, is that jellyfish evolved about 540 million years ago, at more or less the same time as bilaterians. The second discovery is that the jellyfish genome not only includes the same genes that produce the bilateral body plan in flies and fish but also that the genes involved in the production of the bilateral body plan are active during embryonic development in the jellyfish. The genes that produce a front and back during bilateral embryonic development are busy producing proteins during jellyfish development! These discoveries raise all manner of intriguing puzzles, including the question why the bilateral genes in the jellyfish fail to produce a bilateral body plan.

The crucial point for my purposes, however, is that these discoveries concerning the age and constitution of jellyfish genes—paradigmatic, low-level systemic mechanisms—have forced biologists to reconsider their theories of the anatomy of jellyfish. Discoveries at the level of genes have provoked changes in the concepts with which we understand large-scale phenotypic traits. As Zimmer reports, John Finnerty, a biologist at Boston University, predicts that further research will show that the nervous system of the jellyfish, far from being an undifferentiated net, "is divided into specialized regions like the human brain" (Zimmer 2005). We could hardly ask for a more vivid illustration of the power of analyzing inward. Nor could we ask for a clearer demonstration of the wisdom of following the directive in (EC), of calibrating our hunches and intuitions so that we are open to alteration in the categories with which we understand natural systems. The importance of (EC) can, in fact, be put quite generally: the confidence we invest in the concepts with which we understand high-level systemic capacities should always be proportioned to what we actually know of the low-level mechanisms that constitute the system.[3]

It may be worth pointing out that a commitment to the fecundity of analyzing inward is not a commitment to ontological reductionism. This is clear from what we have recently learned about reductionism in biochemistry. Consider the following form of reductionism:

Reductionism (R): For any living system S, all high-level capacities are explicable in terms of generalizations quantifying over intrinsic properties of low-level systemic mechanisms.

The properties of low-level mechanisms are "intrinsic" just in case they do not depend on the mechanisms being part of system S—just in case, that is, they are features that these low-level mechanisms possess even when extracted from S or placed in some other type of system. The key intuition behind (R) is that high-level systemic capacities are wholly explicable in terms of certain core properties of low-level mechanisms; stability in high-level capacities is, presumably, a direct product of stability in the core properties of low-level mechanisms. But we now know that (R) is false, at least for certain kinds of biochemical systems. The work of Boogerd et al. (2005) shows that the behavior of certain complex cells decomposes not simply into the intrinsic causal properties of low-level enzymes but also into properties that enzymes acquire only in so far as they are related in the right way to certain other chemical agents.

Boogerd et al. are clear that the falsity of (R) does not entail the failure of mechanistic explanation. The high-level systemic capacities of the cells they discuss are entirely explicable in terms of the causal and mechanical effects of low-level mechanisms and their relations. Nevertheless, the kind of reductionism expressed in (R) is demonstrably false, since some of the high-level capacities exemplified by living systems are explicable only in terms of the properties of low-level mechanisms that depend crucially on their place in precisely those kinds of living systems. This is enough to show that a commitment to analyzing inward, to explaining high-level capacities in terms of low-level mechanisms and their interactions, in no way entails a commitment to the form of reductionism expressed in (R). It is also worth emphasizing that the discovery reported by Boogerd et al. was made precisely because they approached the study of complex cells by following the lessons that motivate the directive in (EC)—by analyzing inward and synthesizing laterally.

History of Culture

The history of (at least) Western culture should make us self-consciously suspicious of what I am calling dubious concepts. As I note in chapter 2, one category of dubious concepts comprises concepts that descend from a worldview we no longer regard as true or promising. The traditional concept 'free will' as elaborated by Chisholm (1964) and others is dubious in this sense since it derives directly and proximately from our theological ancestry and since it has not been vindicated by any contemporary form of inquiry. And recent theories concerning the apparent purposiveness of living things, discussed in chapters 4 and 5, provide additional illustrations of concepts dubious by descent.

We should not underestimate the staying power of concepts dubious by descent. Such concepts are perpetuated, of course, because our cultural institutions tend to be forces that stabilize and thus conserve. And at least part of the explanation for the conserving powers of our cultural institutions appears to be the effects of certain entrenched features of our psychology. As the scientists Peter Richerson and Robert Boyd (2005) observe, humans are uniquely good imitators, far better than other primates. They draw on work by Michael Tomasello (1999) and others suggesting that very young children are highly successful imitators when compared with our nearest cousins, the chimps. And it is this, according to Richerson and Boyd, our capacity for imitation, which helps ensure the replication of skills, beliefs, values, and attitudes—and, I might add, all the associated conceptual categories—across generations. Replication occurs through myriad channels, including education and mass media, where they compete with one another for our attention and adoption. This in turn helps ensure the selective efficacy at the cultural level of the conceptual categories involved.

Moreover, as the political scientists Pippa Norris and Ronald Inglehart (2004) observe, ancestral social values and their associated conceptual categories tend to linger in a culture long after the ancestral worldview has lost its former dominance. In the course of defending a thesis of secularization—the thesis that there has been a shift in most developed and developing countries away from sacred and toward secular worldviews—they offer evidence that the central categories of our conceptual scheme, including concepts concerning our selves, exhibit a form of inertia that is far stronger than any concern we might have for theoretical coherence: "The second building block for our theory assumes that the distinctive worldviews that were originally linked with religious traditions have shaped the cultures of each nation in an enduring fashion; today, these distinctive values are transmitted to the citizens even if they never set foot in a church, temple or mosque" (Norris and Inglehart 2004, 17). And they give the following illustration: "Thus, although only about 5% of the Swedish public attends church weekly, the Swedish public as a whole manifests a distinctive Protestant value system that they hold in common with the citizens of other historically Protestant societies such as Norway, Denmark, Iceland, Finland, Germany, and the Netherlands. Today, these values are not transmitted primarily by the church, but by the educational system and the mass media" (Norris and Inglehart 2004, 17). Although they put the point in terms of values, the values involved must, of course, be conceptualized in some form or other. Consistent with the thesis of cultural evolution in Richerson and Boyd, the relevant causal forces, according to Norris and Inglehart, include education and the media. And it goes without saying that education and the media can have such effects only for animals endowed with entrenched psychological capacities that result in social learning.

One particularly important point confirmed by the massive data that Norris and Inglehart analyze in their study concerns the possibilities of change across generations. The basic thesis of secularization they defend is that increases in existential security—in security against war, crime, or oppression, in access to adequate food, safe drinking water, education, and so on—tend to cause a decrease in religiosity, measured mainly by the frequency of participation in religious ceremonies and in the degree to which religious values are subjectively important. The point about generations then is that the degree of existential security experienced early in life appears to be decisive. Adults living in countries that experience a sudden increase in existential security tend to retain the degree of religiosity acquired early in their lives; a dramatic change in one's fortunes as an adult has relatively little effect. By contrast, the children of these adults—those born during or after the rise in security—tend to be far less religious than their parents. This suggests that it is a fact of human psychology that the degree of security experienced early in one's life is far more formative than the degree experienced later on.

This is important because it generalizes in several ways. It suggests, for instance, that Darwin, writing in the final chapter of the *Origin,* was quite right to think that his theory would garner widespread support only in future generations. Darwin knew that the theory of evolution by natural selection is a threat to many a mind. So he says:

Any one whose disposition leads him to attach more weight to unexplained difficulties than to the explanation of a certain number of facts will certainly reject my theory. A few naturalists, endowed with much flexibility of mind, and who have already begun to doubt on the immutability of species, may be influenced by this volume; but I look with confidence to the future, to young and rising naturalists, who will be able to view both sides of the question with impartiality. (Darwin 1859, 482)

The important point for my purposes concerns the difficulties we face when trying to adhere to the directive in (DD), given below. It appears that the emotional or aesthetic register with which we respond to the world—the way we respond, for example, to potential threats—is more or less set early in life. Our emotional register is, without doubt, structured by our genetic and developmental inheritance.[4] We may, in consequence, be constitutionally inclined to resist the directive in (DD) when the topic of our inquiry leaves us vulnerable to felt threats. And, as we will see in part 3, recent discoveries in neuroscience and psychology provoke questions concerning the nature of human agency that cut very close to the bone. We may hope, I suppose, that we are endowed with the "flexibility of mind" possessed by a few naturalists known to Darwin, though we should guard against trading wishful thinking for sobering facts.

These reflections on the history of culture and on the causal relation between the conserving effects of cultural institutions and human psychology warrant the following directive:

Concepts dubious by descent (DD): For any concept dubious by virtue of descent, do not make it a condition of adequacy on our philosophical theorizing that we preserve or otherwise "save" that concept; rather, bracket the concept with the expectation that it will be explained away or vindicated as inquiry progresses—as we analyze inwardly and synthesize laterally.

The importance of (DD) consists not simply in a general mistrust of concepts with a theological heritage but also in serving as a check on the intellectual biases bequeathed to us by our predecessors. The hope is that we become self-conscious of the prejudices we may otherwise fail to notice in ourselves. In doing so, we may create the intellectual space within our own inquiries to apply the lessons that motivate the expectation in (EC). We are, after all, the products of our history, including our cultural history, and it is naive to think that we can, simply by way of self-reflection, identify and eradicate the retarding effects of our culturally instituted categories. Perhaps the most pernicious effect of concepts dubious by descent is that we tend not to see the need for, or the possibility of, analyzing inwardly and synthesizing laterally. We tend to see certain concepts as inescapable and constitutive of what we are, even when those concepts are the products of historical and cultural forces that are escapable after all.

Having a dubious history is not, of course, an automatic verdict of guilt; a concept may descend from a worldview we no longer regard as promising and nevertheless refer to a causally efficacious property in the world. So the directive in (DD) is not equivalent to simple eliminativism. The directive is that we bracket concepts dubious by descent until progress in inquiry explains them away or vindicates them. Both possibilities, elimination and vindication, must be left open; that is what a progressive orientation requires. So the aim of (DD) is twofold. One aim is to withhold authority from concepts that are dubious by descent; we should not hold our intellectual ambitions hostage to concepts bequeathed to us by a worldview we no longer accept. But at the same time, the aim is to rest the issue whether to jettison or preserve a given concept on the fruits of relevant inquiry. This too is to subject ourselves to the world.

Human Psychology

Some of our most basic conceptualizing capacities lead us astray. As mentioned, contemporary cognitive psychologists maintain that the human mind comprises a host of conceptual and inferential capacities. Some of these capacities, it is claimed, dispose us to apply the concept 'cause' under a range of conditions, including conditions in which the apparent causal relation does not obtain. Others dispose us to apply one or more of a cluster of mental concepts, including 'intention,' even when the attributed intention is absent. These latter capacities go well beyond the distinctively human facility for imitation. They appear to be highly structured yet deeply entrenched capacities that underwrite our social intelligence. And as we will see beginning in chapter 6, these capacities are prone to generate an abundance of false positives, which renders the concepts involved dubious by virtue of their psychological role.[5]

Now, even if you have a healthy skepticism toward contemporary cognitive psychology, the point here is still significant. It is plausible that some of our most basic cognitive and affective capacities enable us to anticipate and navigate our environments under a limited range of conditions. It is also plausible that we find ourselves, often enough, in conditions outside those limits. The resulting false positives may be, from an evolutionary point of view, a cost worth paying; the false positives may be a nuisance or even deleterious in some instances, but the presumption is that being possessed of different psychological structures would be far worse. However, from the point of view of trying to acquire knowledge, the false positives may be intolerable. As Robert McCauley (2000) suggests, the pursuit of scientific knowledge is probably unnatural—contrary to some elements of our nature—in precisely this sort of way. If so, then progress in knowledge is limited by the retarding effects of our own psychological architecture or, less pessimistically, by the bounds of our best efforts to creatively think or feel our way around our own structural limitations.[6]

If all this is right, then it is crucial from the point of view of inquiry that we become self-conscious about the conditions under which such conceptualizing capacities may lead us astray. We need a directive concerning concepts dubious by psychological role:

Concepts dubious by psychological role (DP): For any concept dubious by psychological role, do not make it a condition of adequacy on our philosophical theorizing that we preserve or otherwise "save" that concept; rather, require that we identify the conditions (if any) under which the concept is correctly applied and withhold antecedent authority from that concept under all other conditions.

As with the directive concerning concepts dubious by descent, the importance of (DP) is that it serves as a check that enables us to better apply the lessons that motivate (EC). It serves as a check against the biases of our own psychological structures; it creates the space, by way of analyzing inward and synthesizing laterally, for us to better understand the limitations of our own abilities.

Of course, (DP) may be difficult to implement. It may be difficult because the capacities engage nonconsciously and because they are entrenched and thus constitutive of the way we orient ourselves to the world. It may also be difficult because the concepts, even if they reach the level of conscious reflection, are central to how we portray ourselves as agents and as inquirers, and they sometimes distort or conceal the truth about ourselves. As we will see (especially in part 3), implementing (DP) may require that we devise strategies for thinking or feeling our way around certain naturally distorting dispositions of thought and feeling.[7] And we must be prepared to discover that implementing this directive is beyond our powers with respect to some psychological capacities.

Evolutionary History

At least two elementary facts concerning our evolutionary history are important. The first is that we are evolved animals. All of our capacities, including our capacities to generate and apply concepts, are products of our evolutionary history. The conceptual schemes we generate are also the products of psychological mechanisms evolved over very long stretches of time. The second elementary fact is that we, along with other primates that exist today, evolved from common hominine[8] ancestors, in which case the study of our cousins is sometimes a crucial aid to the study of ourselves. These facts, coupled with claims from psychology or neuroscience, enable us to formulate reasonable hypotheses concerning the structure of our psychology. They enable us to distinguish structure from noise. They do this by making it rational to choose hypotheses that fit what is known about our evolutionary history over hypotheses that do not.

If taken in isolation, such evolutionary considerations are weak, but if combined with the lessons that motivate the expectation in (EC)—if combined with the fruits of analyzing inward and synthesizing laterally—then considerations of our evolutionary history have force. A plausible evolutionary hypothesis serves as a general framework for the pursuit of low-level mechanisms and interactions; the more plausible the framework, the more compelling the low-level taxonomy. We thus should require the following:

Evolutionary history (EH): For any hypothesis regarding any human capacity, make it a condition of adequacy that, as we analyze inward and synthesize laterally, we do so within a framework informed by relevant considerations of our evolutionary history.

It bears repeating that the directive here is weak. I do not share the confidence of so-called evolutionary psychologists who claim that we can, by assembling broad generalizations concerning our evolutionary history, discover fine-grained facts about the structure of our minds today. I do of course believe that the human mind today is very much a product of our evolutionary history, but I have little confidence that we know enough to make much of this fact in trying to understand the specific structures that constitute our psychology. Such confidence is unwarranted, I think, by the fact that we have far too little reliable information concerning the relevant history. As Robert Richardson (2007) argues to great effect, if evolutionary psychology is to succeed in making psychology a branch of evolutionary theory, it needs to rise to the evidential standards of evolutionary theory, and for the most part it does not—our access to the relevant historical information is too limited.[9] The information that *is* available, however, when combined with the fruits of analyzing inward and synthesizing laterally, can indeed help us distinguish structure from noise. The case of the jellyfish discussed above is illustrative.

So, too, is the following case. Consider a simple behavioral difference between rats and cats. If we electrically stimulate the correct part of a rat's brain, namely, along the lateral hypothalamic corridor, the animal will engage in vigorous exploratory behavior, much as it would when presented with an unfamiliar environment or when foraging for food. If we stimulate the same neural area of a cat, however, the animal does not scurry and sniff, but stalks. Its posture and movements are suddenly transformed into those it employs when preparing to attack. It is tempting to claim that the behavior of these animals is obviously different—the rat is exploring its environment, the cat is aggressive and predatory. But that is misleading if not false. We already know that the anatomy and chemical processes of these animals are highly similar when they behave in these ways. We also know that the two behaviors, despite their apparent differences, are not different at all from an evolutionary point of view. That is, rats and cats are relatively close mammalian cousins with similar neural mechanisms implementing apparently different behaviors. The force of (EH) in this case is to consider whether seeking in rats and stalking in cats serve a common evolutionary function. And indeed they do. Rats are scavengers and eat only by effectively rummaging around; cats are predators and eat by effectively stalking and killing. This, then, is simple illustration in which the common nature of two apparently different behaviors is distinguished from an otherwise noisy scene. It is also a case in which, without the framework imposed by (EH), we misunderstand the very phenomena we wish to explain. The cat's stalking, in the absence of (EH), appears a straightforward form of aggression and not a form of food exploration.[10]

There is, moreover, an aspect of the progressive orientation toward inquiry that is underwritten quite firmly by the directive in (EH). Progressivism requires that we engage in inquiry with the attitudes and expectations with which naturalists of the nineteenth century embarked on their voyages, with the expectation of discoveries bound to unsettle our current worldview. This orientation remains relevant today when the objects of our inquiry are systems we understand little or not at all—our capacities as agents included. When the nature of human agency is the object of our inquiries, the *mere fact* that we are evolved animals ought to engender the expectation that our worldview is bound to change. This is to take a radically historical view of ourselves, as Darwin does in the *Origin* regarding all forms of life, and taking a historical view of ourselves is bound to produce a startling gestalt shift. The conditions under which our ancestors evolved were probably substantially different from conditions in which many humans live today. At least some of our capacities—those descended from such ancestors—may be expressed today only rarely or in ways we find puzzling. Without knowing the historical conditions in which our capacities evolved, we may overlook or misinterpret the systemic functions they serve. This, at any rate, is a possibility we cannot dismiss out of hand. And since we are probably prone to misinterpret our own capacities, we cannot take our initial descriptions of the phenomena we wish to understand at face value. Even the most innocent descriptions of our cognitive and affective capacities cannot be taken as correct or as likely to survive the new knowledge gained as we analyze inward.

Anticipatory Systemic Functions

The fifth and final directive described in this chapter is what I hope is an obvious corollary to the directive in (EH), namely, that we frame our inquiries into our capacities as agents by seeking their anticipatory systemic functions:

Anticipatory systemic function (A): For any psychological capacity of any minded organism, expect that among its most prominent systemic functions is the function of anticipating some feature of the organism's environment.

In all probability, most or all of our affective and cognitive capacities are oriented toward the future, toward what is likely to happen next or relatively soon in one's environment. Why? Because the environments of our ancestors, like those of most living things, no doubt changed in ways that made successful anticipation of those changes utterly vital. This is obvious regarding external conditions such as changes in weather, patterns of predation, rearing of young, and so on. But it is equally obvious regarding changes internal to the organism, such as the pain of hunger or grief. And this makes it highly likely that all, or nearly all, our affective and cognitive capacities are highly oriented toward the organism's future. This even applies to capacities such as memory and dreaming that we typically conceptualize in ways that are not particularly anticipatory (more on those in chapter 6). We may have some psychological capacities that are not anticipatory in this way, but it is prima facie plausible that such capacities are a small minority.

Concepts of Agency

I come now to a preview of the specifics of my discussion. I will be concerned throughout with various concepts clustering around the notion of agency. In part 3, I focus mainly on theories of human agency and thus on concepts that cluster around the notions 'free will' and 'moral responsibility.' In part 2, I focus on theories of life that extend back to the eighteenth century, with emphasis on theories of biological purpose and design from the last few decades. I should like to explain why I proceed in this way.

Our traditional understanding of the nature of life was driven by the observation that living things possess three remarkable capacities: they can incorporate external elements, both living and nonliving, into their own form; they can perpetuate their specific form over time; and they exemplify an extraordinary degree of functional interdependence of parts. None of these capacities belongs to nonliving things. Living things are endowed with the capacity for self-organization and self-perpetuation, while nonliving things are organized or perpetuated by external factors. Or so it seemed. An adequate theory of life, in consequence, was driven to posit some sort of internal, form-giving power to all living things. This, as we will see, is the focus of Kant's discussion of natural purposes in his *Critique of Judgment.* It is also the motivation, as we have already seen, for Goethe's antecedently existing, form-giving archetype. On this view, what distinguishes the living from the nonliving is, in very general terms, the existence of an internal center of command and control. The very shape of life and its exquisite fit to the world are conceptualized as the effects of what had to be a centralized source of power capable of imposing order and functionality. And this power, it turns out, was difficult to understand except on the model of conscious, creative agents like ourselves! Our traditional concept of 'life' drives us inexorably toward concepts of creative agency.

It is useful to begin with this traditional concept of life for several reasons. The first is that tracing the genealogy of 'purpose' or 'function' illustrates one way in which progress in knowledge forces us to alter or eliminate, rather than preserve, what initially seemed to be an ineliminable concept. Our intellectual predecessors, Kant included, thought the concept ineliminable because they could not imagine how to conceptualize living organisms absent an internal center of command and control. The very attempt to do so must have felt deeply unsatisfying, a mere changing of the subject, a bit of sophistry. This appears to have been the attitude of the German Romantic naturalists, Goethe and Humboldt included, all of whom drew inspiration from Kant's *Critique of Judgment.* And yet, as I will describe, progress in scientific knowledge provides us with conceptual tools that Kant and others did not possess and understandably failed to imagine. We now possess tools with which to conceptualize living things as devoid of a center of command, as transient ensembles of fragmented, scattered, and contingent causal factors, as what some have aptly described as tokens in ongoing cycles of contingency.[11] Explicating 'purpose' in terms of agency is no longer a viable option; discoveries in biochemistry and progress in evolutionary theory have put our traditional concept to rest. Conceptual progress of this sort illustrates the power of the progressive orientation toward inquiry. It is in light of this illustration that we should approach the topic of human agency directly in later chapters of the book.

A second reason for beginning with theories of life is to illustrate the extent to which the theologian's orientation persists in the contemporary intellectual landscape. I will focus on its persistence mainly in contemporary philosophical circles, although academic philosophers are by no means unique in this regard. In particular, I will focus on current philosophical theories of biological purposes or functions. All the theories I examine have been offered by philosophers who are, by their own avowals, naturalists in some form or other, and yet, despite these avowals, the theories offered all retain vestiges of the concept of agency to which traditional theories of life appealed. Ruse is explicit that he intends to preserve a version of Kant's view. Other theorists—advocates of the theory of proper functions, in particular—substitute the shaping powers of natural selection for the formative power of an agent, and they offer this as progress. But, as I will argue, it is a mistake to conceptualize natural selection as a surrogate center of command and control. As important as it is to understanding life, not even natural selection acts as a centralized source of power capable of creating and perpetuating forms of life. The forms that exist are products of a host of scattered causal factors; these forms are vulnerable to changes on the outside as much as to changes from within. And it is only because these philosophers retain the orientation of a conceptual conservative that they find these theories of functions the least bit compelling.

The third reason for beginning with theories of life is the striking parallels to theories of human agency. As indicated, traditional theories of life sought to account for the evident differences between the living and the nonliving, between things with the inherent capacity to perpetuate their form and those with no such capacity. Likewise, traditional theories of human agency aspire to explain the evident difference between selves and nonselves or, more precisely, between rational agents and non-rational agents (as well as nonagents). The latter include plants and lowly animals; the former include humans. And just as Kant and his contemporaries were compelled to theorize that living things are animated by a center of command and control, something similar is true of traditional theories of human agency. Indeed, the human agent has been conceptualized as the earthly exemplar of a centralized source of creative and controlling capacities. And yet, just as discoveries in biology tell against such centers of command in the case of living things, discoveries in neuroscience and psychology tell against them in the case of human agency. To the extent we are agents at all, we bear little resemblance to the traditional view of ourselves that runs from Descartes through Kant and well into the twentieth and twenty-first centuries.

Now it may turn out that our understanding of human agency does not go the same way as our understanding of life. And since inquiry for the progressive is exploration, we must leave open the possibility that human agency is an anomaly in the way suggested, for example, by Niebuhr. At the same time, however, we must take seriously the extent to which progress in our knowledge of human nature already suggests a parallel. As I describe in later chapters, our knowledge of our own constitution and our cultural history is progressing much as our knowledge in biology progressed throughout the twentieth century. And since inquiry always begins with a bet on which horse is going to cross the finish line, it is surely irrational to ignore what we already know regarding the actual workings of our capacities as agents.

I wish to emphasize that my thesis concerning the fate of 'human agency' is not nearly as strong as my thesis concerning 'purpose' in biology. The latter concept, on my view, is dead, and it is time that biologists and philosophers faced this fact. My thesis concerning the concepts that cluster around 'human agency' is twofold. First, we do not yet know what kind of agent we in fact are, though we do know that we are not the sorts of agents that our traditional notions of agency suggest. Second, the way to discover what kind of agents we are—our best prospects for success—is to cultivate a progressive orientation by adopting my directives for inquiry. That, I think, is where we are at this point in the history of inquiry—our concepts of agency are in transition, we have no clear idea what we are, yet we know enough to judge that our best bet is to bury the orientation of conservatives and imperialists and cultivate the progressive orientation of the naturalist-as-explorer.

Finally, at key points in my discussion I lapse into polemics on methodology. Among the most central tasks, perhaps *the* central task, of contemporary philosophy is what I am calling the concept location project. As described in chapter 2, the aim is to preserve or otherwise "save" concepts of apparent importance by locating them amid the concepts and claims of some preferred theory. Frank Jackson (1998), who aspires to what he calls "serious metaphysics," gives preference to theories in ontology. For naturalists, the preference goes elsewhere, usually to some theory in science or to some set of methods of inquiry culled from this or that science. In either case, however, the task is the same: try to preserve what appears worth preserving by placing it within a theoretical landscape that is deemed antecedently unobjectionable. Philosophical debate, on this model, reduces to claims and counterclaims over which philosophical theory succeeds in preserving the greatest number of apparently important concepts. Thus, nonnaturalists like Jackson argue for a dualist ontology on the grounds that a purely physicalist ontology fails to "save" certain concepts concerning consciousness, while so-called naturalists defend a variety of strategies, all of which hinge on the question whether a given concept can be, or should be, preserved within a naturalistic ontology or within naturalistic methods.

We may leave aside the details that divide the philosophical terrain, for this entire form of "inquiry" is intellectually bankrupt, no less so than the reflections that Niebuhr offers in place of genuine inquiry. There is, in the end, almost nothing of substance separating naturalists of our day from their alleged antithesis, the nonnaturalists of our day. Both parties perpetuate the theologian's orientation by adopting a conservative or imperialist attitude toward concepts of apparent importance. This is evident in many ways, including, for example, the assumption that an adequate philosophical theory is one that preserves our core conceptual intuitions concerning the relevant phenomena. This, as we have seen, is the assumption that drove Chisholm concerning 'free will.' It is also, as we will see, the same assumption driving the theories of natural purposes and the theories of human agency that I discuss below. If we learn anything from the study of our predecessors, we ought to learn that the preservation of concepts dubious by descent or by psychological role is a recipe for theology and, in consequence, the obituary of genuine inquiry. We can no longer take it for granted that preserving our conceptual intuitions is a mark of success. To the contrary, it is a reliable rule of thumb that the more a theory preserves our intuitions regarding a dubious concept, the more likely it is leading us away from the truth. Our methods of philosophical inquiry are in need of reform. They need to be purged of the vestiges of our long theological ancestry that still beat deep in our hearts.[12]

2

The Allure of Agency

'Purpose' in Biology

FOUR

The Real Heart of Darwinian Evolutionary Biology

The metaphor of design, with the organism as artifact, is at the heart of Darwinian evolutionary biology. **MICHAEL RUSE, *DARWIN AND DESIGN***

With one exception, all contemporary theories of biological functions or natural purposes are unacceptably conservative or imperialistic. A deeply conservative orientation is part of the very framework of Ruse's (2003) defense of a neo-Kantian metaphor of design. A similar orientation frames the theory of proper functions defended by a small army of theorists over the last thirty years or so. The main source of their conservatism is the stubborn insistence that an adequate theory must preserve the alleged normativity of functional traits. This is to endorse the default assumption that the alleged content of apparently important concepts must be preserved. The lone exception to such conservatism is the theory of systemic functions articulated by Robert Cummins (1975) and developed by Ron Amundson and George Lauder (1994) and also by myself (Davies 2001). On this view, our understanding of functions must be reformed to better fit a progressive orientation toward inquiry, and that requires giving up our traditionally normative understanding of functions.

The aim of my discussion throughout part 2 is twofold. I wish to convince you that our traditional notion of normative functions is indeed something we must put behind us. We now know too much about the theological origins of that concept and about the actual workings of living things to grant it any authority in our inquiries. We also know that we are psychologically inclined to conceptualize objects in our environment as purposive even when they are devoid of such purposes. The notion of normative functions is a dead concept, and it is time we recognized it as such. Part of my argument is that the cost of retaining the normative concept of functions involves postulating a prior agent of some sort or other. Our theologically derived concept of natural purposes is retained only if we accept the existence of a center of command and control endowed with agent-like powers. This, however, is no longer an acceptable price to pay because our best scientific theories tell against it. It is time to jettison the concept as construed in our theological ancestry and embrace the theory of systemic functions.

The second aim is to convince you that the directives for inquiry described throughout part 2 are among our best reasons for rejecting the traditional notion of normative functions. The aim is to demonstrate the power of the progressive orientation by way of historical illustration, by way of progress in knowledge that is difficult to contest. And, having demonstrated the power of my directives in our understanding of natural purposes, it then becomes plausible to expect similar results in our understanding of human agency. The more general aim of part 2, at any rate, is to set the stage for the application of my directives in part 3.

The focus of the present chapter is Ruse's metaphor of design. I argue that the metaphor ought to be rejected in light of the dubious history of the concepts involved. This is clear in light of the directive concerning conceptual change in (EC) and especially the directive concerning concepts dubious by descent in (DD). The focus of chapter 5 is the theory of proper functions, which is, I argue, even less naturalistic than Ruse's metaphor. The theory gives the appearance of being naturalistic because it appears to rest directly on the theory of evolution by natural selection. But that appearance is a facade. As I will show, the theory flouts not only the directives in (DD) and (ED) but even the directive in (EH) that we frame our inquiries in light of what is known about our evolutionary history. Finally, in chapter 6, I argue that both views—Ruse's metaphor and the theory of proper functions—run afoul of the directive in (DP) concerning concepts dubious by psychological role.

The Metaphor of Design

In *Darwin and Design,* historian and philosopher of science Michael Ruse defends a theory of biological design that is vaguely Kantian in content. The theory purports to explain in naturalistic terms the design that living things appear to exhibit, including the apparent purposiveness of their parts. He frames his discussion in terms of a puzzle. Most everyone familiar with evolutionary theory agrees that the argument from design, popular in Britain before and during Darwin's time, was defeated by Darwin's discovery. That is how Darwin came to see matters. In his *Autobiography* he says, "The old argument of design in nature, as given by Paley, which formerly seemed to me so conclusive, fails, now that the law of natural selection has been discovered.... There seems to be no more design in the variability of organic beings and in the action of natural selection, than in the course which the wind blows" (Darwin 1969, 87).

The theory of evolution by natural selection explains the diversity and adaptedness of living forms better than any form of theology. Darwin, in effect, killed design. At the same time, evolutionary biologists today talk rather freely of functions and purposes in the course of theorizing about the living realm. Thinkers most intimately acquainted with evolutionary theory talk without blushing as if the very thing Darwin had killed were alive and flourishing in the biological sciences. So what is going on?

Ruse's solution to the puzzle is conciliatory, a fact nicely reflected in his title, *Darwin and Design.* Darwin killed design—that much is beyond dispute.[1] And although Darwin's murderous ways show that there is no literal design in the living realm, there nevertheless is, according to Ruse, an indispensable metaphorical role filled by the concept of design. The practice and indeed the very existence of evolutionary theory, he claims, require the metaphor. The reconciliation thus consists in seeing with one eye that there is no biological design while directing the other eye to view living things as if they were designed. The first eye keeps it sights on the truth, but those truths are illuminated and visible only if the second eye manages to "see" what is in fact not there. The concept of design is reconciled with Darwinism because 'design' is a metaphor we Darwinians cannot do without.

In defending this reconciliation, Ruse offers the crucial premise that while there is no literal design in the living realm, the metaphor of design is practically necessary for theorizing in evolutionary biology. He offers two general considerations in support of this premise: first, the apparent necessity of the metaphor induced from a handful of evolutionary studies and, second, the role that 'design' and 'function' have played historically, especially in biology, natural theology, and philosophy.

Upon reflection, however, the premise that the metaphor of design is practically necessary for evolutionary theorizing cannot be sustained. Ruse's first argument is easily defeated by the simple fact that we have an alternative theory of biological functions, one that preserves everything of value within evolutionary theory but denies the existence of design (literal or metaphorical) among living organisms. Or so I shall argue. I shall then argue that Ruse's second argument is defeated in light of Darwin's rhetorical insights, especially his caution concerning culturally inherited habits of thought. Indeed, the historical considerations marshaled by Ruse in defense of his metaphor—the first nine chapters of his book trace the historical roots of our concept of normative functions, including the rather fat roots that run through the natural theology of the eighteenth and early nineteenth centuries—are precisely the sorts of considerations that tell *against,* not for, the metaphor. The concepts 'design' and 'purpose' are, as Ruse himself demonstrates, dubious by descent. To insist that the history of our concept serves as evidence that the concept must be preserved is a particularly perverse form of conceptual conservatism.[2]

Ruse's Reconciliation

Ruse's first premise is that the mechanisms of natural selection gives rise to apparently organized, complex adaptations. Adaptations are descendents of selectively successful lineages. The human thumb is an adaptation on the assumption that ancestors with thumbs produced (in the long run) more offspring than ancestors with functionally different thumbs or with no thumbs at all. Some adaptations, moreover, in the process of being selected for, acquire a striking degree of internal complexity and organization. Indeed, Ruse assumes that selection is the only natural process likely to produce organized complexity. This is not to assume that all adaptations are complex or organized—that is false, as melanism in the peppered moth shows—but rather that all or most organized, complex traits are adaptations.[3] The peculiarly shaped lens in the trilobite eye (more on that presently) provides a striking illustration.

The second premise is that organized, complex adaptations strike us as analogous to artifacts. This is a thesis in human psychology. It is natural for human inquirers to conceptualize the components of living organisms the same way we conceptualize artifacts. The analogy hinges on the property of end-directedness, in this sense: artifacts contribute to human interests or ends, including those of the designer, producer, or consumer, and, by analogy, a complex adaptation contributes to the organism's survival or reproduction, which we may conceptualize as organismic "interests" or "ends." The claim then is that the causal contributions of an adaptation to the "ends" of survival and reproduction strike us as directed in the same way as the causal effects of artifacts:

We are faced [in the case of living organisms] with what we have been calling, in as neutral a way as possible, "organized complexity." ... Whatever you call it, this complexity allows for and indeed calls for understanding in human terms of intentionality, of purpose, of *design*. Organized complexity is artifactual. That was the whole point of natural theology.... Whether or not organisms really are designed, thanks to natural selection, they (or rather, they inasmuch as they are adaptive or adapted) seem as if designed (for the ends of survival and reproduction). We may no longer be thinking of a literal designer up there in the sky, but the mode of understanding persists. For the natural theologian [William Paley, for instance], the heart is literally designed by God—metaphorically, we compare it to a pump made by humans. For the Darwinian, the heart is made through natural selection, but we continue, metaphorically, to understand it as made by humans. (Ruse 2003, 265)

The workings of complex adaptations make it natural for us to view them as purposive relative to the ends of survival and reproduction, even while believing they are not purposive at all.

The third premise is that the exercise and existence of evolutionary theory require that we embrace our natural tendency to see living things as if they were artifacts. Human beings would not theorize about the evolution of living organisms without conceptualizing organisms as end-directed. The claim, more specifically, is that absent the metaphor we would fail to ask questions or formulate hypotheses that throw light on the causes of evolution. We would fail to ask the "why?" and "what for?" questions that fuel the discovery of evolutionary histories. The entire science would be extinguished:

If we want a biology that is not interested in the reason why the stegosaurus has such a funny display on its back, is not intrigued by the peculiar shape of the trilobite lens, does not care why some butterflies mimic other butterflies ... then presumably something can be done. Some low-grade biological activities like classification, perhaps, can indeed be done better than usual if we refrain from asking why ... I presume we could also do a certain amount of embryology and physiology and other sorts of biological activity without any questions about adaptation.... But it is all surely going to be very limited.... Without the metaphor, the science [of evolutionary biology, at minimum] would grind to a halt, if indeed it even got started. The plates are there; the lens is there; the mimics are there .... Brute statements, which probably we would not be led to make in the first place because there would be no reason to even notice. Who would care about the trilobite lens? Why bother to describe it rather than something else? (Ruse 2003, 284–85)

This premise, like the previous one, is a thesis in psychology. Absent the metaphor, it is unlikely that observations of complex, organized traits would incline us to ask why they evolved or what they evolved for. We might make lists of traits, including those that are complex and organized, but we would not be moved to inquire after the causal powers that led to their selective success.

The fourth and final premise is that, on pain of destroying evolutionary biology, we must embrace the metaphor of design. We cannot help it anyway, so why resist? Resistance would only thwart the progress of evolutionary theory. We ought to embrace Darwin and the death of literal design, to be sure, but in embracing Darwin we should also employ, perhaps even cultivate, the metaphor of design. Any other stance amounts to a denial of the kind of inquirer we are.

Metaphor or Systemic Functions?

The third premise that the metaphor of design is necessary for evolutionary biology is crucial to Ruse's argument. But is it true? Ruse's first line of evidence appeals to the process of reasoning employed by evolutionary biologists in a handful of specific cases. In each case, biologists begin by noticing a trait that is complex or organized in some way and, on the assumption that selection is the most likely cause of organized complexity, they work to discover the selective pressures and mechanisms that led to the evolution of the trait. It is here that the metaphor of design comes to the fore. The claim is that, in order to inquire into the selective history of a complex and organized trait, biologists must conceptualize the trait as engineers would an unfamiliar artifact. They must conceptualize the trait as if designed and constructed for some end. Evolutionary inquiry, at least for organized, complex adaptations, is structured by asking "why?" and "for what?"

Consider the case of the trilobite eye, which contains a double lens divided by a peculiarly shaped barrier. It appears natural to wonder why there are two lenses, the upper lens and the lower intralensar bowl? And why the peculiarly shaped barrier? It turns out that, if the organism had only the upper lens, its visual system would be unable to produce clear images of perceived objects. The upper lens has a refractive index that produces spherical aberration, causing a point object to be perceived as a blurred disk. The intralensar bowl has a refractive index that, when joined to the upper lens by the peculiarly shaped interface, corrects the aberration. The result is a visual system that produces clear, sharp images. What is wonderfully striking about all this, as Ruse points out, is that the aberration is corrected in precisely the way described in the optical theories of Descartes and Huygens, theories based on the properties of artificially produced lenses. Both theorists identified the refractive index required to correct the aberration produced by a simple lens. We have, then, a striking coincidence: optical theories of an artifact—beginning with a simple lens—apply exactly to the evolved visual system of the trilobite. "Such a coincidence," according to Ruse, "is no chance but evidence that something is afoot. More precisely, the coincidence points to a design-like feature of the organic world, which Darwinians take as evidence of adaptation—features possessed by organisms because they help in the struggle to survive and reproduce" (Ruse 2003, 266). The salient conclusion is that, in trying to understand why the eye of the trilobite contains an intralensar bowl in addition to an upper lens, theorists unavoidably turn to considerations of engineering. They unavoidably conceptualize the trilobite eye in the same way Descartes and Huygens conceptualized an artifact.

Ruse offers a few other cases and then generalizes. In each instance, the evolutionary theorist, in order to make progress in understanding a trait's selective past, employs the engineering strategy and thereby conceptualizes the trait as an artifact. In making that move, the theorist does not forget that Darwin killed design. The engineering move is construed metaphorically, not literally. That, at any rate, is Ruse's recommendation concerning contemporary talk of natural purposes.

But is the premise true? No, it is not true. Its falsity, in fact, is easily demonstrated. I grant that, in theorizing about the trilobite eye, evolutionary biologists (a) ask "why?" and "what for?" questions, (b) attribute functions and purposes, and (c) appeal to engineering models. Still, it does not follow that evolutionary biologists must employ the metaphor of design in order to theorize as they do. It does not even follow that biologists in fact employ the metaphor. That would follow only if there were no alternative theory of biological functions capable of sustaining such theoretical activity. The simple fact, however, is that there is such an alternative.

The alternative is the theory of systemic capacity functions. Ruse mentions the theory only to set it aside, on the grounds that its original formulation was not intended to apply to adaptations (Ruse 2003, 260–62). This, at any rate, is the charge leveled against Cummins (1975, 1977, 1983), who developed the theory in the context of psychology, and against Amundson and Lauder (1994), who developed the theory in the context of anatomy and morphology. However, Ruse's attempt to neutralize the theory in this way has no force. It may be true that Cummins makes rather little of the applicability of the theory to adaptations, but the relevant question is whether the resources of the theory allow for such applications. I have argued in Davies (2000, 2001) that they do. The claim is that the application of Cummins's theory to populations that evolve or remain in equilibrium via selection is all that evolutionary theorists need by way of a theory of functions. This is no small point in assessing Ruse's reconciliation.

The domain of the theory of systemic functions is the set of all systems, natural and artificial. It applies to assembly lines that produce automobiles but also to the capacity of any population to evolve or remain in equilibrium via natural selection, as well as the capacity of, say, the trilobite eye to focus light and produce clear images. To apply the theory, we (1) isolate the system we wish to study, (2) specify the high-level capacity of the system we wish to understand or control, and (3) analyze into the system and identify low-level mechanisms that, acting in concert, produce the specified high-level capacity. Having accomplished that, we may (4) attribute to low-level mechanisms the systemic function of producing precisely the causal effects identified in step 3. To illustrate, we isolate the automobile assembly line, specify the high-level capacity we wish to understand (producing *n* automobiles in a given period of time, for example), and analyze the line into its structural and interactive components until we identify the causal effects of each component integral to the specified high-level capacity. We may then attribute systemic functions accordingly. And this strategy iterates inward. We may be especially interested in a specific component of the system—the component, say, that preassembles the motor. Having isolated that component and its high-level capacity, we apply steps 1–4 yet again and analyze into the parts of the subsystem that assemble motors. We repeat these steps still further to analyze a high-level capacity of a motor part, a high-level capacity of parts of motor parts, and so on.

It is significant that, at any level of analysis, we may ask of some systemic component, What is it for? We may ask, that is, what a component is for relative to our systemic analysis of the specified high-level capacity. It is likewise significant that our analyses not only permit such questions, they answer them as well. We can ask, for instance, of any component of the assembly line what it is for and, once we apply the strategy in 1–3, we arrive at our answer in step 4. The systemic function, for example, of this particular component relative to the high-level capacity of producing automobiles is the preassembly of the motor.

Crucially, the very same questions can be asked and answered for complex, organized adaptations. Consider again the trilobite's eye. In Riccardo Levi-Setti's marvelous discussion—the source of Ruse's discussion—we are presented with extraordinary photos of trilobite fossils, including several detailed depictions of the structure of the eye (Levi-Setti 1993). It is clear from the fossils that the structure of the intralensar bowl corrects for spherical aberration precisely as recommended by Descartes and Huygens. Moreover, even if Descartes and Huygens had never formulated their optical theories, it is nonetheless plausible, given the precision and complexity of the trilobite eye, that some other enterprising scientist would have discovered that these features of the trilobite eye mimic those of artificially produced lenses. At any rate, we apply the theory of systemic functions to this case by identifying the proximate system—the visual system—and high-level capacity—the capacity to produce clear images. We analyze into the system and quickly hypothesize that the salient systemic effect of the intralensar bowl is the correction of spherical aberration, which confers upon the combined lenses the systemic effect of producing clear images.

This, however, does not yet establish that the double lens is an adaptation; our attributions of systemic functions are not yet attributions of adaptive systemic functions. To show that the intralensar bowl was selectively successful by correcting spherical aberration and that the double lens was successful by producing clear images, we need to locate the visual system and the organism in a larger system relevant to the process of natural selection. That, of course, is the population. We begin by assuming the existence of an ancestral population containing trilobites with the peculiarly shaped double lens and trilobites with a variant lens. The population is fruitfully conceptualized as a system, and the high-level capacity we wish to explain is the ability of the population to increase the relative frequency of the peculiarly shaped double lens. Steps 1 and 2 are thus fulfilled. We satisfy step 3 by discovering the selectively salient systemic components. For simplicity, let us assume that during the relevant period of selection there was competition between just two groups of organisms in the population, those with the newly mutated, peculiarly shaped double lens and those with some variant. These two groups, being the relevant units of selection, constitute the salient systemic components for this system. Each component contributed uniquely to the relative increase in the frequency of the peculiarly shaped lens—organisms with the double lens reproduced more frequently or with greater long-term viability than those without it.

Having isolated these components and contributions, we may then analyze inward by way of steps 1–4. We must extend the strategy in this way if we wish to discover the systemic functions of traits at the level of the trilobite eye. At this point in our reasoning, we return to our initial systemic function attribution, namely, the function of correcting spherical aberration. Now, however, our function attributions have been placed in the context of an adaptationist hypothesis: the intralensar bowl, by correcting the spherical aberration produced by the upper lens, contributed to the efficacy of the double lens as a whole, which contributed to the efficacy of the visual system, which contributed to the organism's capacity to survive, which contributed to relatively greater chances for reproduction. Precisely this hypothesis explains the increase over two or more generations in the relative frequency of the double lens.

Crucially, as in the case of the assembly line, we can ask of the intralensar bowl or the double lens, What is it for? This question is best understood as shorthand for the more informative, By virtue of what causal contributions did ancestral tokens of this trait contribute to the selective efficacy of the type? In asking "what for?" we are not asking for some end or purpose, metaphorical or otherwise. We all agree that Darwin killed design. What we do ask—and what gets us into the *real* heart of Darwinian evolutionary biology—is the question, What systemic function was fulfilled by ancestral tokens of the intralensar bowl and double lens such that this type of eye was selected for? Or, by virtue of what causal powers did ancestral tokens confer on future generations the status of being an adaptation? The theory of systemic functions, therefore, generates all the relevant questions and hypotheses integral to evolutionary theorizing. Nothing is lost, save the excesses of Ruse's metaphor.

It is significant that the theory of systemic functions is far more general than Ruse's metaphor, which applies only with respect to the "interests" of survival and reproduction. Unlike Ruse's metaphor, steps 1–4 apply not only to traits that are selected for but also to traits selected against. Here is a fanciful illustration. Until recently, it was widely believed that the enormous tusk of the male narwhal whale played some role in sexual selection.[4] And this bit of fiction made it easy to imagine ways in which the oversized tusk, which sometimes grows to nine feet in length, could readily be selected against. An increase in tusk size, we might suppose, was initially selected for by virtue of arousing the interest of females. But as its size increased further, its effectiveness diminished. It grew unwieldy, reducing the animal's speed and coordination, resulting in more and more lost battles, fewer and fewer reproductive opportunities. If we consider the period of selection in which the oversized tusk is selected against, the relevant capacity of the population is the ability to increase the relative frequency of the smaller tusk. And during this period the cumbersome tusk clearly acquires the systemic function of reducing the organism's effectiveness in combat—simply apply the strategic steps 1–4 if you are skeptical—even though whales with the unwieldy tusks suffered relatively greater injury and death. The "ends" of survival and reproduction were positively thwarted, in which case the metaphor of design does not apply to the tusk, yet the attribution of a systemic function is informative and justified by the conditions of the theory.

The theory of systemic functions, then, in explaining how a population evolves, justifies the attribution of functions to all causally efficacious components of the population, to types of organisms selected for and types selected against. In addition to explaining how a population changes, however, the theory can also explain how a population remains the same, how it remains in equilibrium. Imagine an ecological niche in which two competing species manage to coexist over several generations in relatively stable proportions, thereby violating the *competitive exclusion principle.*[5] This can happen for different reasons, but the most likely explanation is that something in the shared ecology differentiates between the two species despite their overall similarities. It may be, for example, that species A better exploits a food source during winter months and species B does better during summer months. If some such recurring change in the niche differentiates between species A and B in this way, it may result in a cyclical equilibrium. And we can explain all this by applying the theory of systemic functions. We first conceptualize the two species in a common niche as the larger system and then identify the relevant high-level capacity, namely, the capacity of the system to remain in equilibrium over several generations. Steps 1 and 2 are thus satisfied. Then we analyze into the system and individuate its causally efficacious components: we isolate the causal differences between the two species, namely, the capacity of A to better survive winter months and the capacity of B to better survive summer months. If we wish, we may then analyze into species A to explain its capacity to better survive winter months, enabling us to identify systemic functions of causally efficacious parts of organisms in species A. And so on for species B.

Above I conceded that evolutionary theorists, in the course of theorizing about the selective history of some complex, organized trait, do indeed (a) ask "why?" and "what for?" questions, (b) attribute functions and purposes, and (c) employ engineering models. All three practices are explicable on the theory of systemic functions, though a few nips and tucks are called for. The practice in (a) is ambiguous between actually employing the metaphor of design (as Ruse recommends) and merely asking "what for?" as shorthand for, By virtue of what causal contributions did ancestral tokens of this trait contribute to the selective efficacy of the type? (as I recommend). Now, if any biologists today in fact employ the metaphor, then, in the interests of saying what is true, they ought to stop it. As Ruse insists, Darwin killed design; that is something that evolutionary biologists know to be so. Why persist in saying what we know to be false or concealing what we know to be true? We have, after all, an alternative theory of functions that better reflects what we know. If, on the other hand, biologists ask "what for?" as mere shorthand, then the theory of systemic functions applies full force and the practice in (a) can be preserved. The practice in (b) can be preserved too, if we take the suggestion of Amundson and Lauder (1994) and talk more precisely about systemic capacities without literal or metaphorical ends. Finally, (c) requires a slight but revealing gestalt shift. When evolutionary biologists appeal to engineering models—as Levi-Setti allegedly does—the stress may fall upon the fact that the model depicts an artifact, but it may fall somewhere else. It may fall not on the depiction of an artifact but rather on the depiction of a system comprising two or more levels of organization. And insofar as the theory of systemic functions is plausible, that is how we should interpret (c). Of course, many artifacts are systems of the relevant sort, which explains why artifacts are often helpful in uncovering the causal effects by which ancestral tokens were selectively successful. But there is no requirement that the model depict an artifact. The only requirement is that it depict a system, artifactual or otherwise. In consequence, most of the practices described in (a) – (c), appropriately revised, are explicable from within the theory of systemic functions.

I conclude, therefore, that Ruse's third premise is false. The metaphor of design is practically necessary only if there are no other practical options. But there are others (at least one), so the alleged necessity does not obtain. I also conclude that my alternative to Ruse's metaphor is superior on the grounds that it better reflects what we know to be true without hindering the practices of evolutionary theorists in any way. This, I believe, is argument enough that Ruse's metaphor is antithetical to the orientation of the progressive naturalist.

The Relevance of History

If the above conclusion is correct—if the existence and fecundity of an alternative theory of functions defeats Ruse's third premise—we should be puzzled. Why does he endorse a premise so easily defeated? Why, indeed, when the alternative theory is clearly known to him? The answer, I think, is simple. In Ruse's judgment, and in the judgment of most theorists, the fatal flaw in the theory of systemic functions is that it makes no attempt to preserve the alleged normative or value-laden dimension in the content of our concept. And that, it is commonly assumed, renders the theory irrelevant to evolutionary theory:

There does seem to be something distinctive about biological understanding—something having to do with purposes and ends in evolution. It seems, at one level, to lie in the fact that in evolutionary biology we are dealing with peculiarly connected, causal relationships. We have "final causes." One set of things causes another set of things, but the first set in turn seems to exist *because* it brings about the second set. But more: the second set of things is in some sense desirable, to be valued. (Ruse 2003, 268)

That a theory of biological functions or purposes must retain this appeal to values is revealed, according to Ruse, by the very same considerations that support this third premise, namely, the actual practice of evolutionary theorizing and the long, persistent history of the concepts: "Both history and present Darwinian evolutionary practice have shown us that this kind of design-type thinking [which appeals unavoidably to norms or values] is involved in the adaptationist paradigm" (Ruse 2003, 268).

But this line of reasoning is objectionably conservative. The appeal to values in "design-type thinking" provides not the slightest reason to turn our backs on the theory of systemic functions. Two points are relevant. First, considerations from the previous section suffice to show that the appeal to evolutionary practice cuts no ice. It is simply not true that we need to posit values, norms, or end-directedness of any form—literal or metaphorical—in order to ask and answer the questions that drive the construction of evolutionary hypotheses. The theory of systemic functions enables us to ask "why?" and "what for?" and provides a general strategy for answering these sorts of questions without the attribution of natural purposes. Second, the strategy Ruse uses to marginalize the theory of systemic functions—his claim that the theory fails because it excludes any element of end-directedness or all reference to values—reveals the depth of his commitment to a conservative orientation. Of all the elements contained in the putative content of our traditional concepts 'design' and 'purpose,' those that seem to require the existence of an end or a norm of evaluation are surely closest to the theological worldview we no longer accept. Ruse is arguing, in effect, that the theory of systemic functions is irrelevant to understanding living things because it is too distant from the theological worldview that Darwin's theory so roundly defeated! The conservative affinities between Ruse and Niebuhr are, at this point, impossible to ignore. Why else insist on preserving a concept that flaunts its theological pedigree, except out of fervent allegiance to conceptual conservatism? In contrast, it is precisely the distance between our ancestral worldview and the theory of systemic functions that recommends this account of functions to the progressive naturalist.

If the actual practices of contemporary evolutionary biologists fail to support Ruse's metaphor, then what are we to make of his other line of argument concerning the long, persistent history of 'design' and 'purpose'? If, as I have argued, the theory of systemic functions is superior to the metaphor of design, what weight can the history of our concepts carry? This question is best answered in light of my directives for inquiry. Or so I wish to argue. I will focus my discussion on four directives for inquiry, and I will try to convince you that, in light of these directives, Ruse's metaphor of design is something we should reject even if we lacked a better alternative—even if the theory of systemic functions had not been discovered—on the grounds that the metaphor is at odds with the directives described.

It is perhaps worth noting that my directives help us discover not only how to frame our intellectual tasks but also what our intellectual problems are. It is naive to think we need only look around us to discern the problems that impede progress in knowledge. Acquiring a feel for real problems requires a great deal of knowledge. The mere fact that a given concept or question has persisted for a long time is by no means sufficient, since we all know how and why young scholars perpetuate the categories and problems of their teachers and journal editors. We also know how difficult it is to keep up with progress in other areas of inquiry and to know when or how far such progress has undermined our current conceptual scheme. So knowing *that* an apparent problem really *is* a problem is a challenge unto itself, and part of the significance of my directives is that they help point us toward features of our history that bring into focus impediments to progress. One way they help is by making clear which concepts from our intellectual history are prone to retard rather than facilitate our efforts.

My first directive has an air of such simplicity that it may strike you as utterly pedantic:

Descriptive accuracy (DA): When inquiring about any phenomena, identify the target of our investigation as fully as, but not more fully than, our initial grasp of the phenomena allows.

Upon inspection, however, this directive is not so simple. In some cases it is terribly difficult to honor. The difficulty is most acute for phenomena we think or feel we already understand. Especially when the object of our inquiry is ourselves, it is tempting to think or feel that we are already endowed with at least some substantial knowledge of what we are like. It is tempting, for example, to feel that the concepts in terms of which we understand ourselves as agents are uncontroversial—we feel we have a first-person intimacy regarding the content of these concepts—and this only strengthens our conviction that these concepts rightfully serve as the basic framework for our investigation. Once we give in to this temptation, we quite naturally become hardened in our pretheoretical expectations and hunches. We may allow minor adjustments to our concepts of, say, free will or moral responsibility, but substantive changes now appear tantamount to changing the subject. Or so it feels.

And it is precisely this feeling that makes the directive concerning descriptive accuracy especially pressing when the objects of our investigations are most likely to inspire a conservative orientation. (DA) is an expression of caution that applies to naturalists and nonnaturalists alike.

The second directive is the following:

Theoretical competition (TC): On the basis of the description of the phenomena that best adheres to (DA), develop alternative and competing theories of those phenomena and devise experiments to discover which theories are most predictive and explanatory.

We have, in the case of living organisms, only one conception ready at hand. We cannot, with William Paley, conceive of organisms as literally designed because we are assuming that Darwin killed design.[6] We are also assuming that the theory of systemic functions has not yet been discovered. That leaves us with Ruse's metaphor. We should, therefore, on the basis of (TC), strive to formulate an alternative conception of living things absent the metaphor of design. This is so, at any rate, on the assumption that inquiry is most fruitful when theories have to compete with one another (more on this below).

The third directive is the expectation in (EC) of conceptual change. As we analyze inward and synthesize laterally, especially for systems we understand poorly or not at all, we should expect that the concepts in terms of which we understand the high-level capacities of the system are likely to be altered or abandoned. This is an informed expectation. The history of science makes it rational to expect that increases in our knowledge will force revisions in the categories with which we conceptualize high-level systemic capacities. Recent discoveries in science have a similar effect. If our molecular biology conflicts with our chemistry, or our geology with our physics, something must change. And if, as in the case of jellyfish, our knowledge in genetics forces us to revise our theories in embryology, anatomy, and evolutionary history, that is all to the good.[7]

Returning then to Ruse's assertion that the metaphor of design is practically necessary for evolutionary theorizing, perhaps the first thing to notice is that we should feel strong pressure to formulate a concept of biological functions without the metaphor of design. We may find ourselves drawn to the metaphor of design—we may even consider ourselves unable to theorize without it—still, the metaphor is false and we know it. So the joint force of (TC) and (EC) in this case is considerable. We should at least attempt to formulate an alternative conception of functions, one that coheres with the natural sciences generally but also with the specific theory it is supposed to serve. And we should do so expecting that our core conceptual categories will be changed as a result. To think otherwise—to insist on a metaphorical version of a concept bequeathed to us by our theological ancestors—is to close off the possibility that we can create new concepts that enable us to better grasp the way the world is.

Of course, we may fail in our attempts to formulate an alternative theory of functions, but that is not the point. The point is the method—a fallible method fallibly induced from the history of science. At the heart of this and other naturalistic methods is the assumption that, as we make progress in inquiry, the goal is to conform ourselves to the world. This more than anything else is the mark of a progressive naturalist. The working expectation is that inquiry into the world will change our concepts and beliefs, our emotional and aesthetic sensibilities, with the cumulative effect that our thoughts and feelings better fit the way the world is. The aim is not to isolate concepts near to our hearts and insulate them from revision, but to articulate our concepts and make them vulnerable. Vulnerability to revision and loss is our best hope for fitting our concepts to the way the world is, and since no other method of inquiry or reflection has yet to match the mechanisms of correction that characterize the inductive sciences, a scientific orientation with an explicit openness to conceptual change is by far the best horse on which to bet. Precisely this is the argument for setting our concepts in competition, so we may discover which ones best cohere with the concepts and claims of other sciences and which ones survive the discovery of low-level systemic mechanisms.

Ruse may object that I have misconstrued his appeal to metaphor. He says more than once that metaphors—extensions of concepts from one domain to some other—are of great heuristic value in science. He also notes that, while the application of a metaphor is always false (since the concept being extended has its home in some other domain), truth is not the purpose. The heuristic value of metaphors consists in causing us to see things in ways we otherwise would not see them, to ask questions or formulate hypotheses that otherwise would not occur to us. This, he insists, is integral to progress in science. He even appeals to Darwin's use of metaphor to defend the integrity of his own reconciliation: "Here I stand with Charles Darwin" (Ruse 2003, 284).

But we should not be fooled by these claims. The general claim that metaphors assist in scientific discovery is one we can accept, at least for the sake of argument. The pertinent question is whether the conditions under which metaphors usually contribute to discovery in science obtain in the case of biological design, and obtain in such a way that the metaphor is rendered necessary to the very existence of evolutionary theorizing. It is hard to believe that they do. Darwin, after all, borrowed the concept 'selection' from animal and plant breeders—from its application to intentional (that is, mindful) processes—and extended it to nonintentional, natural forces that produce a struggle for existence. His use thus qualifies as metaphorical. However, in extending that concept to the struggle for existence, Darwin does not suggest that the metaphor is necessary to understanding his theory. He does not assert, for example, that we must pretend to see the natural forces that give rise to the struggle for existence as if those forces were intentional in some way. More generally, in applying the metaphor of selection, Darwin does not suggest that we believe or pretend to believe something we know to be false. Ruse, by contrast, wants us to believe in the practical necessity of a metaphor we know to be false. Worse, he wants us to believe in the necessity of a metaphor within evolutionary theory, even though the metaphor is undermined by precisely that theory. It is hard to see how, in this instance, Ruse is standing with Darwin.

There is the further point that, in extending the concept of selection, Darwin was not preserving it, but changing it. He extended, broadened, and thereby altered an otherwise homely concept. He borrowed a bit of conceptual material already in our repertoire and put it to a new and constructive use. Darwin's extension of the concept 'selection' was creative. Ruse's metaphor, by contrast, is not creative in this way, but conservative. Ruse recommends that we preserve the core content of our concept and merely alter our attitude toward that content. Instead of the attitude of belief—instead of asserting that living things really are designed—it is recommended that we adopt an attitude of pretense—pretending that living things are designed. This is conceptual conservatism in action. It is hard to believe that Darwin would have endorsed the use of metaphor in the service of conservatism. He surely would have refused a metaphor that his own theory discredits.

What then is the value of history for philosophical naturalism? Ruse seems to think that the history of the relevant concepts supports the practical necessity of the metaphor of design. How? Perhaps by its sheer persistence, or perhaps by its (alleged) transformation from a theologically based concept to one that appears explicable in terms of natural selection. But neither point is plausible. Even on the assumption that the theory of systemic functions has not yet been discovered, the methods expressed in (TC) and (EC) should drive us to look for an alternative. Those methods provide the impetus for further inquiry that would likely lead us away from the metaphor and toward the formulation of a theory that, at minimum, does not conflict with or conceal the implications of evolutionary theory. More generally, the directives in (TC) and (EC) express a spirit of openness to the world, of putting our concepts and claims on the line, and thereby conforming our ways of thinking and feeling to better fit the world.

And this brings me to a further constraint on naturalistic methods, namely, the directive in (DD) that we withhold antecedent authority from concepts dubious by descent. This is a directive induced not from the history of science specifically but rather from the history of culture. The idea is that some of our concepts, in light of their cultural genealogy, ought to be regarded as suspect for purposes of inquiry. Some concepts—those concerning the nature of the self, most obviously, but also concepts of purpose and design—have a dubious history. A history is dubious if, for example, the concepts historically had their home in a largely theological context we now regard as false. As I have said, this is not to assume that all theological theories are false, only that none has shown itself relevant to the enormous progress in knowledge since the rise of modern science. Besides concepts with overt theological ancestry, additional categories of dubious concepts can be identified by appeal to other substantive or methodological claims in our ancestral worldview that we now believe (with relevant justification) to be false or irrelevant. Concepts that presuppose the existence of Cartesian nonphysical substances are another example. The notion of a nonphysical soul is dubious because it conflicts with the assumption that the world is physically closed, and this assumption is warranted by the extraordinary predictive and explanatory successes of our best-confirmed physical theories. Neuroscientists, for instance, do not require of their theories that they somehow preserve or account for our concept of the soul. Why should philosophers proceed differently?

At a minimum, we should not assume that a successful theory of the relevant domain must in some way "preserve" or "account for" dubious concepts. To do so is to fail to see where we are in the history of inquiry: we know too much to take seriously any attempt to do physics or chemistry or biology or psychology on the basis of substantive theological assumptions. Why should philosophy be any different in this regard? Why should we take seriously any philosophical discussion that begins with the assumption that we must preserve or otherwise account for our concept of design?

(DD) may appear to conflict with my earlier claim that naturalism at its heart involves an openness to the world. Withholding antecedent authority to dubious concepts, it might be thought, exemplifies a closing off, not an opening up. There is, however, no conflict here. The main reason that concepts dubious by descent are dubious—the reason, for example, that theological notions have played a diminishing role in the progress of knowledge—is precisely because *they* have the effect of stifling inquiry. This is clear from the fact that theological notions tend to run afoul of the strategies that motivate (EC). They do not cohere well with the concepts and claims of our well-developed sciences and, as Darwin complained in the *Origin,* it is difficult to analyze into any natural system on the basis of theological concepts. The creationists of Darwin's day appealed at crucial points to the creative powers of some divinity but consistently failed to specify the secondary (that is, earthly) mechanisms involved. The alleged theological action always occurred in the absence of mechanisms accessible via the experimental method. That, of course, made it difficult to confirm or disconfirm any theological hypothesis in terms of low-level mechanisms. The general point then is that theological concepts and claims have the effect of insulating themselves from the directives expressed in (TC) and (EC), directives that have proven immensely fruitful in all other contexts. That is why the directive in (DD) to neutralize dubious concepts that tend to violate (TC) and (EC) is very much in the spirit of opening ourselves up to the world.

The same point holds for philosophical theorizing. I suspect, for example, that the directive against concepts dubious by descent is enough to scotch Ruse's metaphor. Not that I doubt Ruse's intentions. In a spirit of earnest conciliation, he wants to give what he regards as due weight to our theologically derived concept of design. He wants to do what most philosophers do—"save the phenomena," especially concepts that strike us as integral to the problem at hand. But that is a mistake. Some of our concepts are clearly dubious, the concept of design included. To set for ourselves the intellectual task of reconciling what we know to be true (evolutionary theory) with a concept we know to be dubious (design, including metaphorical design) is, when seen from a distance, a nonstarter. Or at least it should be a nonstarter. It is a waste of resources, for one thing, but it also fails to take account of where we are in the history of inquiry. As I say, no one would set for chemists or physicists or biologists or psychologists a research program based on substantive theological concepts or claims. Why should philosophers be any different? Why accept the task of contemporary philosophers—the task of trying to integrate traditional humanistic concepts with all that is currently known about the universe—when, in order to preserve concepts of apparent importance, we end up turning our backs on so much of what is currently known?

Conclusion

According to the interpretation of Darwin provided in chapter 1, we are subjects of the world naturally biased toward our cultural heritage. The conceptual categories of our intellectual predecessors tend to influence the orientation toward inquiry that we take for granted. Yet when the roots of these categories trace more or less directly to a worldview we now regard as false, those concepts are likely to thwart our efforts at inquiry. How could they not retard our efforts, given that they have their home in a view of the world we no longer endorse?

Darwin's rhetorical strategy was to anticipate that certain concepts at the heart of the creationist worldview—concepts that informed the orientation of nearly all his readers—were likely to blind his readers to the power of his theory and negatively color their visceral and intuitive responses. His insight was to anticipate that concepts dubious by descent would cause his readers to recoil from his view of life. My claim in this chapter is that Darwin's insight, applied to Ruse's theory of biological design, reveals a bias in Ruse's orientation toward inquiry akin to that of Darwin's creationist peers. Ruse even tips his hat to the theologians of the eighteenth and nineteenth centuries, the very theorists against whom Darwin was arguing:

The pre-Darwinian philosophers saw the problems. They did not have the naturalistic answer that Darwin gave to us, and for this reason their solutions could be only partial .... The natural theologians downplayed or had difficulties with the nonadaptive aspects of life (such as homology). But the positive overrides the negative. *Where we are today fits very comfortably with the Western tradition of thinking about organisms and final causes. Ours is an evolution from the past rather than anything radically new. And that in itself is a comforting conclusion.* (Ruse 2003, 270; my italics)

These are the sentiments of an unfettered conceptual conservative. It may be true that where we are today fits our traditional ways of thinking about organisms and final causes. But if so, then the pursuit of knowledge has fallen into a degraded state. Just as the production of new life necessitates the incessant destruction of life, so the drive toward greater knowledge necessitates the destruction of past conceptual categories. The thought that the concepts of our theological ancestors ought to be preserved, when seen from the point of view of a progressive orientation, is itself a thought that ought to be eradicated. It is far too pernicious to be retained. Inquiry is for explorers, not taxidermists. It is a sure bet that we have turned our backs on genuine inquiry when we find ourselves gauging the goodness of our conclusions by the degree of comfort they provide.

At any rate, being suspicious of concepts with a reasonably clear theological ancestry—withholding antecedent authority to those concepts—is one way to subject ourselves to the world. We see how far we can progress by withholding authority to a concept that is difficult to synthesize with the concepts of other sciences and difficult to sustain as we analyze inward and uncover low-level mechanisms. It is an interesting and important question just how many alleged philosophical problems of our day would vanish or transform themselves into something more enlightening were we to develop and apply more widely the directive in (DD) against concepts dubious by descent.

FIVE

A Formative Power of a Self-Propagating Kind: Natural Purposes and the Concept Location Project

An organized being is then not a mere machine, for that has merely *moving* power, but it possesses in itself *formative* power of a self-propagating kind which it communicates to its materials though they have it not of themselves; it organizes them, in fact, and this cannot be explained by the mere mechanical faculty of motion. **IMMANUEL KANT, *CRITIQUE OF JUDGMENT***

Conceptual conservatism, as I say, comes in two distinct species. The first species, the commitment to preserve concepts that appear important within some well-developed scientific theory, is illustrated by Ruse's metaphor of design—as are the retarding effects of that commitment. The second form of conservatism includes a commitment to preserve as far as possible concepts that appear important within our general worldview. Many naturalists are committed to this broader form of conservatism. They aim to preserve apparently important concepts that presently play no definite role in any science by locating them amid the concepts or claims of some well-developed science. This is to engage in what I am calling the *concept location project,* the attempt to integrate apparently important concepts from our humanistic tradition with the concepts and claims of our best sciences. The strategic assumption is that, by integrating our nonscientific and scientific concepts, we thereby reconcile and perhaps unify our view of human nature with our emerging scientific view of the world. An aspiring naturalist could hardly ask for more. Or so it seems.

The project of concept location is ubiquitous among academic philosophers. 'Beauty' is a notion that once played a prominent role in our broader worldview and aestheticians with naturalistic sensibilities endeavored to locate this or related concepts in theories of human physiology, psychology, or sociology. The location strategy is similarly employed for other apparently important concepts, including 'knowledge,' 'reason,' 'consciousness,' 'value,' 'freedom,' and more. The ubiquity of this strategy is easy to understand. Since the rise of modern science and certainly since the discoveries of Darwin and Wallace we have become ever so vexed to understand what we are. We struggle to know how to see and feel ourselves in light of our evolutionary history and our relatedness to other animals. What better way to bring the nature of ourselves into focus than by integrating our scientific knowledge of ourselves with concepts at the core of human agency? What better strategy for the philosophical naturalist to embrace?

Concepts clustering around the notion of human agency are the topic of discussion in part 3. Here I linger a little longer over the concepts 'purpose' and 'design' as they apply to living things, to further illustrate the power of the directives employed in chapter 4. Our concept 'function,' after all, appears a powerful notion not merely within evolutionary theory but also within our broader worldview. We talk freely of the function of the eye, the heart, and so on, and when faced with an incapacitated token of an apparently functional type, we describe it as *mal* functioning, as failing to do what it is *supposed to* do. Our general concept 'function' appears to include a built-in standard or norm against which token performances are properly evaluated. Some philosophers with clearly conservative instincts find it altogether fitting to try to locate this general, normative concept within the theory of evolution by natural selection. These are the defenders of so-called "proper" functions, and their theory is the focus of this chapter.

As we will see, the theory of proper functions is not for the ontologically reticent. It posits without fear the existence of literal norms of performance; it commits us to the real existence of normative functional properties. I begin by drawing attention to this feature of the theory and by noting how the theory is supposed to represent a naturalistic advance over Kant's account of self-propagating traits. I then argue that, despite its naturalistic aspirations, the theory wantonly flouts my directives for inquiry and in consequence is unacceptably nonnaturalistic. I defend this claim in two ways. First, it is precisely because the theory aspires to preserve a concept of normative functions that the expectation of conceptual change in (EC) is so baldly ignored. Second, the theory also violates the directive in (DD) that we withhold authority from concepts dubious by descent. This becomes clear once we sketch a partial genealogy of the concept 'self-perpetuation,' a concept at the intuitive heart of the theory of proper functions, and observe a fatal equivocation in the theory's application of that concept. Finally, having revealed the failed aspirations of the theory of proper functions, I generalize. Since there is nothing parochial about our concept of normative functions, the failure of the concept location project regarding this concept, I suggest, provides good grounds for rejecting the concept location project generally. Turning our backs on concept location for concepts dubious by descent is an important step in becoming better subjects of the world.

Norms of Performance

In *Darwin and Design,* Ruse does not assert the literal existence of functional norms of performance, only the necessity of pretending that such norms exist. This is clear from the third premise of his argument that the practice of evolutionary theory would grind to a halt without the metaphor of design. But the alleged necessity of pretending that functional norms are real is also evident in Ruse's second premise. There he asserts that we are psychologically inclined to see living systems "as if" designed by an intelligent mind. The force of this premise is that we cannot help but conceptualize living organisms as endowed with the ends of survival and reproduction—just as we naturally conceptualize artifacts as endowed with various ends—and, furthermore, we cannot help but conceptualize the parts of organisms as endowed with purposes relative to the ends of survival and reproduction—just as we naturally conceptualize the parts of artifacts as having functions relative to their ends. We are led by the very constitution of our psychology, according to Ruse, to see living systems "as if" designed by an intelligent mind.[1]

Ruse's view, however, is an anomaly in the current literature. By far the most prominent theory of biological functions is the theory of proper functions. On this view, traits of living organisms can and should be conceptualized in terms of norms of performance, although not for Ruse's reasons. We are right to conceptualize token traits as normative, according to this theory, not because they satisfy some psychological inclination or because evolutionary theory would otherwise wither and die, but for the far lovelier reason that these norms literally exist. Norms of performance are real properties really possessed by real organisms—so says the theory of proper functions. Norms of performance accrue to contemporary tokens of organismic traits so long as they descend from a lineage that enjoyed ancestral selective success. If the four-chambered heart of human beings exists today because ancestral tokens were selectively successful by virtue of pumping in such and such a manner, then token hearts today have the property of being supposed to perform exactly that task.

This is not to speak metaphorically. To the contrary, the functional norm is utterly real. The claim is that each and every human heart today possesses a specific standard or norm, imposed by the selective success of ancestral token hearts, against which its performance is properly evaluated. If the token is operative and executes the same systemic job that made ancestral tokens successful, then it is doing what it is supposed to do. If the token falls short of what led to selective success for ancestral tokens, then it is failing to do what it is supposed to do. And it is crucial that the token retains its functional standing even when incapacitated and unable to do what it is supposed to do. Indeed, according to the theory, the only way to make sense of saying that the token is failing to do what it is supposed to do is to assume that there really is something it is supposed to do and that it is supposed to do it even when it cannot do it. The functional norm, on this view, is distinct from the actual capacity of the token trait. It is precisely this—the literal existence of a distinct norm of performance—that makes the theory of selected functions so audacious.

The crucial question, however, is this: can the concept of literal functional norms applied to organismic traits be located amid the concepts and claims of the theory of evolution by natural selection? If the answer is affirmative, then advocates of proper functions will have scored a significant methodological point. They will have demonstrated the fecundity of concept location with respect to a concept that is clearly dubious by descent. The possibility of such success, however, is a double-edged sword, for if the answer is negative, then, since there appears to be nothing parochial about our concept of normative functions, we should conclude more generally that the concept location project is, from the point of view of acquiring knowledge, a barren method of inquiry when applied to dubious concepts.

Darwinian Selection and Kantian Self-Propagation

The theory of proper functions draws inspiration from an intuition that runs deep in our intellectual tradition. The idea is that functions emerge in the context of systems that are *self-perpetuating,* in systems that contribute in some regular fashion to the maintenance and proliferation of their own form. This insight lies at the heart of the theory of proper functions, where self-perpetuation is fleshed in terms of selective success.[2] Selectively successful organisms perpetuate themselves individually by maintaining systemic integrity and surviving external threats, and they perpetuate themselves as a species via reproduction. The theory of evolution by natural selection, it seems, lays the foundation on which the self-perpetuation of living forms is explicable in purely natural terms.

The echoes of Kant are hard to miss. Kant would agree with advocates of proper functions that organisms are self-perpetuating and that selectively successful organisms perpetuate their own form. The distinguishing feature of living things, according to *Critique of Judgment,* is that they, unlike nonliving things, are *self-organized* and *self-organizing.* Kant extracts the notion of self-perpetuation from the following three capacities of living things:

1. Individual reproduction—the maintenance of life by taking in external elements and incorporating them into the individual's form
2. Generic reproduction—the maintenance of the species in the production of more individuals of that specific form
3. Internal cohesiveness

These capacities, according to Kant, are distinctive of the living; all living things and no nonliving things have all three. These capacities amount to the generic capacity of any living organism to simultaneously perpetuate its self (token) and its species (type).

Although the agreement with Kant is hard to miss, the disagreements are similarly obvious. For Kant, the self-perpetuating capacity distinctive of all life is, at least to the human mind, inexplicable in causal or mechanical terms. And for that very reason we cannot posit the ontological reality of natural purposes. We must settle instead for what Kant calls a *regulative maxim* (as opposed to a *determinant judgment)* concerning natural purposes. We must settle, that is, for an understanding of functions that is heuristic rather than substantive. Instead of asserting the literal existence of natural purposes, Kant insists that we must conceptualize living things *as if* animated by such purposes. The sort of purpose he has in mind—a purpose sufficient to account for all three capacities in 1–3—is an intrinsic formative power that cannot be explicated causally or mechanically and that is best approximated by the kind of design wrought by intelligent minds. All this is accomplished, on Kant's view, by conceptualizing life *as if* created by a designing, intelligent mind.

So two features of Kant's view contrast sharply with the theory of proper functions. For Kant, (1) "natural purpose" is construed heuristically and hence nonliterally, and (2) at the heart of this heuristic is a commitment to conceptualizing living things as animated by an intrinsic power that is noncausal and nonmechanical. Advocates of proper functions reject both features. They reject feature 2 because we *can* explain the capacities distinctive to all of life by appeal to causes and mechanisms postulated in our best contemporary sciences, and we can do this even while retaining the concept 'self-perpetuation.' They reject feature 1 by insisting that our ontological commitments should be governed by the postulations of our best sciences and not by the critical project in which Kant was engaged. That is, for most contemporary naturalists, the Kantian critique of our mental powers—the attempt to discern the limits of thought and, in consequence, the proper constraints on our ontological pretensions—is replaced by the ontological implications of our best-developed sciences. On this view, if evolutionary theory posits (or otherwise commits us to) the existence of functional offices or roles, then we should be literalists about natural purposes.

In its day, Kant's formative power may have felt compulsory given the remarkable nature of the phenomena, but today, in light of progress in the life sciences, the compulsion no longer exists. In the competition among ideas, Kant's formative power has been routed by the growth of scientific knowledge. This is not to deny that there is something distinctive about living things; the phenomena are as remarkable for us as they were for Kant. But contemporary theories, unlike those of Kant's day, *can* explain the differences between the living and nonliving in causal, mechanical terms. They can do this, moreover, minus the postulation of Kant's formative power.[3] And it is this appeal to progress in science that gives confidence to advocates of proper functions. This is why they feel entitled to insist that our concept of natural purposes has been securely located amid the concepts and claims of evolutionary theory. They see themselves as making genuine progress in our understanding of natural purposes by preserving the notion of self-perpetuation in evolutionary theory. Philosophers committed to the location project could hardly ask for more. Or so it may appear.

The Failure of Proper Functions

Such appearances, however, deceive. The naturalistic credentials of the theory of proper functions are fraudulent; it takes only a bit of historical insight to reveal the deceit. Instead of saving the general concept of functional norms, as the theory of proper functions aspires to do, we should bury it. This is clear in detail and in general, in light of what we already know about our history and our constitution.

Details first. The philosopher Ruth Millikan, a lead apostle of proper functions, asserts that "being preceded by the right kind of history is *sufficient* to set the norms that determine purposiveness; the dispositions themselves are not necessary to purposiveness" (Millikan 1989, 299).[4] The claim is that a token trait today, even when incapacitated and unable to fulfill the associated functional task, nevertheless possesses a functional norm imposed by ancestral selective success. Possession of a proper function thus involves more than descent from past selective success. It requires purposiveness distinct from actual capacities. It requires an abstract, non-causal norm of performance produced by the self-perpetuating capacities of one's ancestors. And this, as we will now see, conflicts with minimal constraints on a naturalistic orientation.[5]

To illustrate the theory, suppose we know the following about the mammalian eye:

1. Ancestral tokens enabled organisms to see with such and such a degree of veridicality.
2. By enabling organisms to see in this way, ancestral tokens enabled organisms to better satisfy some demand of the selective environment.
3. By enabling organisms to better satisfy some demand, ancestral tokens enabled organisms to outreproduce (in the long run) competitors not endowed with this type of eye.
4. By outreproducing competitors, ancestral tokens contributed to the self-perpetuation of ancestral organisms and, in consequence, the perpetuation of the lineage of that type of eye.

This, however, is not enough to underwrite the claim that a token retains its functional status in the face of physical incapacitation. Conditions 1–4 qualify the eye as an evolutionary adaptation, to be sure, since all four are required for evolution by natural selection. But nothing in the definition of "adaptation" warrants the attribution of norms of appraisal, and it is crucial to see why this is so.

In evolutionary theory, "adaptation" is defined in purely historical terms. Present-day tokens of a trait qualify as adaptations for some effect if and only if they descend from a lineage of ancestral tokens that were selectively successful by virtue of producing that very effect. And that means that, by definition, having the property "being an adaptation" tells us nothing about the other properties of present-day tokens. Or, more modestly, the most it tells us is that similarly structured organisms in similar ecologies will likely evolve in ways that converge on a similar functional property. The ubiquity of convergent evolution as described by Morris (2003) makes this a plausible inference.[6] But any other inference from "being an adaptation" is unjustified. Knowing that present-day tokens are adaptations tells us nothing about, say, their present utility; they may be adaptations in light of their history but maladaptive in light of their present situation. Similarly, being an adaptation tells us nothing about the normative properties of present-day tokens, since "being descended from selectively successful ancestors" is distinct from "being supposed to produce such and such an effect." Success in the past is not equivalent to possessing a standard of self-evaluation today. What one's ancestors did in the past tells us nothing about what one is supposed to do now or in the future.

Conditions 1–4, therefore, are necessary but not sufficient for proper functions. The attribution of such norms of performance is warranted only if, in addition to 1–4,

5. Present token eyes, which exist today by virtue of descent from self-perpetuating ancestors, acquired the office or role of enabling present-day organisms to see in such and such a way.

Only the addition of condition 5—only the positing of literal offices or roles to perform the relevant task—transforms mere adaptations into traits endowed with functional norms of performance. Advocates of selected functions are committed to this whether or not they realize it, for something has to serve as the naturalistic source for the alleged standard of evaluation, and it is clear that nothing in the biological term "being an adaptation" can suffice. What must be added is the claim that ancestral selective success not only explains why present-day tokens exist but also that such success produces a literal office, a literal norm of performance, that can survive the loss of the causal powers required to discharge that office, and that it bequeaths this office or role to all descendent tokens.

But now the problem for the theory of proper functions comes to the fore, because the office or role in condition 5 flouts even the most modest of naturalistic constraints. Consider first the strategy of analyzing inward. The mechanisms involved in natural selection are causal and mechanical; the office or role required in 5 is neither. The office is identical to possession of a norm of performance that, by hypothesis, persists in the face of physical incapacitation, even when the required physical base for fulfilling the specified functional task is destroyed. The norm, in consequence, is not identical to any physically or causally specified property.

We may come at the point this way. Advocates of proper functions ask us to suppose that ancestral selective success—the past perpetuation of the organism's form—somehow creates and imposes a norm of performance upon all descendents. We are to suppose that the norm is not identical to the actual capacities of token descendents, for otherwise malfunctions would be impossible. We are also to suppose that the norm is not identical to the ancestral selective history itself, since "having ancestors that did F" is not equivalent to "being supposed to do F." So the obvious question is, how do we analyze inward *from* functional norms of performance *to* the mechanisms of selection? How do we analyze inward from the posited norm to those mechanisms that, according to the theory, give rise to such norms?

The answer is clear: nothing in the theory of evolution by natural selection warrants the attribution of noncausal, abstract norms. Why? Because the processes that constitute natural selection, all of which are causal and mechanical, lack the resources with which to generate and impose noncausal, nonmechanical properties. And if something in the theory of evolution by natural selection did warrant the attribution of noncausal, nonmechanical properties, then that would tell against the claim that the theory of proper functions represents a naturalistic advance on Kant's mysterious formative power. The theory, therefore, fails to analyze inward.

Some theorists may resist. They may insist that "having the selectively successful history of doing F" *is* equivalent to "being supposed to do F." They may even insist that the latter is defined in terms of the former and that that is exactly what their theory asserts. This seems to be the strategy Fred Dretske (1995) employs in defending his theory of consciousness. But none of this is plausible. No naturalist worth her salt would claim that the mere act of defining the normative in terms of the nonnormative constitutes progress. The real question for the naturalist, in light of the lessons that motivate the expectation of conceptual change in (EC), is whether anything in the natural world is capable of producing such norms of performance. The question is whether the physical mechanisms that constitute natural selection contain the resources to generate and impose noncausal, nonmechanical norms. And as I have said, when we turn to the world—when we turn to the mechanisms that the theory of evolution by natural selection posits—it is clear that nothing among those mechanisms can give rise to abstract, noncausal norms of appraisal. (If I am wrong about this, it should be easy to refute me: Show me the mechanisms!) To claim otherwise is to express a wish, perhaps, but one for which we have not the slightest evidence.

Consider now the strategy of synthesizing laterally. The theory of proper functions fails to synthesize laterally for the simple reason that it fails to analyze into the theory of evolution by natural selection. The theory of proper functions cannot ride the coattails of a scientific theory to which it bears no substantive relationship. Of course, the naturalistic credentials of evolutionary theory are difficult to dispute, since it synthesizes across a wide range of theories, but the theory of proper functions fails to analyze into the actual mechanisms of selection and hence cannot claim to synthesize in the same way. In fact, the theory of proper functions attempts to graft functional norms onto the theory of selection but fails to specify the natural mechanisms of selection capable of producing such norms. And the failure to specify any such mechanisms is easily explained: *there are none.* The prospects of synthesizing across other, related theories do not exist.

Advocates of proper functions have missed this point. They have claimed that the attribution of normative functions helps render the claims of anatomists and morphologists coherent with the claims of evolutionary theory. The key idea is that relatively broad functional categories such as "being a heart" or "being a circulator of blood" have their source in the theory of evolution by natural selection but range across claims in morphology. The intuition is easy to illustrate. Some hearts are two chambered, some are three chambered, and some are four chambered, yet all belong to the category "hearts," all are classified in functional terms as "circulators of blood." This, it is claimed, shows that the theory of proper functions does indeed synthesize laterally.[7] But the argument is easy to challenge. We can grant that theories in morphology employ functional categories that abstract away from morphological details even while rejecting proper functions. All we need are systemic functions and a bit of induction.[8] Imagine that we analyze into one type of organism and, after isolating the circulatory system, isolate the main engine of circulation. We then attribute to that component the systemic function of circulating blood. Then, of course, we may analyze into other types of organisms and proceed in the same way. Imagine that, in so doing, we begin to discover a range of morphologically distinct mechanisms across a range of different systems that all serve as engines of circulation. We will certainly note the differences in internal structure, but we may also note that all of these mechanisms share the same systemic function relative to our various systemic analyses. Imagine too, just for good measure, that all these systems are related to each other by descent. What, in this case, is a good naturalist to conclude? Surely we will sum up in our minds the similarities in systemic functions observed in our comparative systemic analyses. We can do this even while denying the existence of functional norms that persist in the face of physical incapacitation.

Now for the more general sense in which the theory of proper functions is nonnaturalistic. Begin with the observation that the attribution of functional norms makes sense only in the context of an organism with known ends. This, as we have seen, is how Ruse describes our psychology. We cannot help but conceive of organisms as systems that, by analogy to artifacts, are endowed with the "ends" of survival and reproduction. Advocates of proper functions agree with Ruse that function attributions require organismic ends, but they do not share Ruse's ontological reticence. They audaciously assert that ancestral selective success endows organisms not with metaphorical ends but with the literal ends of survival and reproduction. Having rejected Paley's extrinsic ends (the ends of God) and Kant's heuristic intrinsic ends (the as-if ends of a noncausal, nonmechanical formative power), the theory of proper functions offers instead the literal ends created and imposed by ancestral selective success.

But the key premise is false. The theory of selected functions cannot deliver a third source of organismic ends because the mechanisms of selection are not capable of creating or imposing such ends. To think otherwise is to confuse producing some regular effect with having the end of producing that effect. Organisms with relatively superior causal capacities do indeed survive longer and reproduce more often, on average, than other organisms, and this leads to selective success. But the mere fact that superior survival and reproduction lead to selective success does not entail that superior survival and reproduction are organismic ends. Certain organismic effects are causally necessary for selective success, but having as ends the performance of those effects is not necessary for selective success. Success in selection, though important, is not magic; it cannot transform regular organismic effects into organismic ends.[9] After all, some organisms have capacities that are relatively causally inferior, and, in the right environment, those with inferior capacities reproduce less often and eventually die off. We are not tempted, however, to attribute the end of going extinct to those with inferior abilities, even though that *is* what they are doing and even though, as Darwin insists in chapter III of the *Origin,* the relentless process of extinction is absolutely integral to the struggle for existence and to evolution by natural selection. So why think that those who enjoy selective success are any different with respect to ends? The difference between the winners and the losers is that, in the particular selective environment in which they live, the causal effects of the winners happen to better fit the bill. There is nothing normative in so brutal a fact of life.

Selective success, therefore, is not so wondrous as advocates of proper functions would have us believe.[10] Evolutionary theory describes causal processes and mechanisms that help explain the perpetuation of some forms of life and the eventual failure of all forms of life. That the theory accomplishes this without postulating anything akin to an intrinsic formative power is grounds for rejecting Kant's theory of natural purposes. But it is also grounds for rejecting any sort of surrogate for Kant's formative power. We must reject the claim that selective success produces noncausal, nonmechanical norms of performance. The processes and mechanisms of selection cannot produce such norms, and we know it. The theory of proper functions, therefore, is not a naturalistic advance on Kant's teleology; it is, instead, an instance of the same failed view, disguised in terms borrowed from a scientific theory to which it bears no real relationship.

Trial Balloons and Concept Location

John Post (2006) offers an alternative defense of proper functions that appears compelling on its face. Inspired by David Papineau's 2001 model of reductionism, Post suggests we naturalize our concept of normative functions by reducing it to conditions 1–4 listed above, the conditions involved in descent from ancestral selective success. What is crucial to Post's use of Papineau's model is that reduction proceeds by first formulating a hypothesis that certain phenomena are equivalent to specified low-level mechanisms and then studying the actual world for evidence that confirms and evidence that disconfirms our proposed reduction. If empirical investigation reveals that our hypothesis is predictive and explanatory, we must take it seriously; if not, we discard it. In this way, we study the actual world first and sort our concepts later in light of how the world actually is. This appears to heed the lessons that motivate the directive in (EC).

Consider an analogy to a phenomenon known as "ball lightning"—a term used to describe "fantastic, glowing, floating balls of colored light, often accompanied by a hissing sound and distinct odor" observed by ordinary folk since antiquity (Post 2006, 7). The actual existence of this phenomenon remains uncertain—we have only the reports of the folk—yet scientists have offered theories aimed at explaining its nature. If ball lightning is real, then, according to these theories, it is one and the same as high-density plasma in some form or other. We have, therefore, the trial balloon hypothesis that the remarkable phenomenon reported by the folk is reducible to—identical with or somehow determined by—high-density plasma. Post puts it this way:

BL: *x* is ball lightning if and only if *x* is a high-density plasma of kind *k.* (Post 2006, 11)

This is Post's illustration of Papineau's model of reduction. BL asserts a specific ontological relation (equivalence) between two apparently distinct phenomena (ball lightning and high-density plasma), and the stated equivalence functions as a trial hypothesis. We adopt it and then put it to the test. We see how far it contributes to progress in knowledge and then tidy our conceptual categories in light of such progress.

Advocates of proper functions, according to Post, can apply the same model of reduction to the alleged normative functions that evidently populate the realm of living things and derive precisely the same sorts of theoretical gains. We may adopt—as a trial balloon hypothesis, as an equivalence to be put to the test—the following reduction of normative functions to the process of descent from selective success:

DFOR: *A* is directly *for E* (normative sense) if and only if *E* is the effect of *A′* s past instances in virtue of which *A* was selected for [that is, selectively successful]. (Post 2006, 15)

Like BL, the equivalence in DFOR is put forth not as an analysis of the concept 'purpose' or 'normative function' but as a straightforward, provisional hypothesis intended to guide inquiry. If adopting this hypothesis yields theoretical and experimental fruit, then, according to Post, that is all the argument we need for thinking that DFOR is indeed true. At least one of the lessons that motivate the directive in (EC) appears to be satisfied.

Post does not claim to demonstrate the truth of DFOR. He claims rather that it is rational to accept DFOR in order to see what follows from its adoption. This is rational because the basic model of reduction is one we readily accept elsewhere, with respect to ball lightning and high-density plasma, for example, but also with respect to water and H_2O, and so on. And of course the reason we accept these other cases is because doing so has a proven track record in the history of modern science. We make progress in understanding natural systems by floating these sorts of hypothetical equivalences and then using them to guide us in future inquiries. The value of Papineau's model of reduction is something we glean by paying attention to the history of human inquiry and, as I say, this line of reasoning, at least when seen from a distance, appears to satisfy the lessons that motivate the directive in (EC).

Post's general argumentative strategy is to show that the adoption of DFOR is immune to a host of objections usually aimed at attempts to naturalize an alleged normative property. He argues, for example, that the adoption of DFOR survives the challenges raised by Hume's law argument (that no norm can be inferred from any facts), Mackie's queerness argument (that normative properties are too mysterious to be naturalized), Moore's open question argument (that no norm can be reduced to facts), and so on. And the arguments Post offers in defense of this premise, although by no means immune to challenge, are indeed powerful. If we grant him the premise that DFOR is on a par with BL and with other sorts of hypothetical reductions, then defenders of Hume's law argument, Mackie's queerness argument, and so on have their work cut out for them.

The obvious response to Post's argument, however, has nothing to do with the usual litany of objections that he discusses. It has to do rather with the directives for inquiry I have described. To see this, begin with the question, Is it true that we must accept the proposed equivalence in DFOR in order to "see what follows?" On the assumption that we accept analogous hypotheses in the cases of ball lightning and water and so on, is it the case, as Post alleges, that the refusal to accept the hypothesis in DFOR would amount to an objectionable double standard?

No, it is not the case, as the lessons that motivate (EC) make plain. We know from the history of modern science that progress in understanding natural systems requires that we analyze inward to low-level mechanisms and processes. It thus is rational to refuse any proposed equivalence when the prospects for analyzing inward do not exist. The adoption of Post's DFOR is rational only if we have been given reasonable grounds for thinking that we can analyze inward *from* the normative properties associated with proper functions *to* the mechanisms the constitute ancestral selective success. But as I say, not a single advocate of proper functions has succeeded in identifying mechanisms of natural selection capable of producing and imposing noncausal norms of evaluation, and it is difficult to see where they might look for help. Post might reply that we should keep an open mind and allow that biologists may, in the future, discover such mechanisms, but that would be disingenuous. The theory of evolution by natural selection as presently understood is not subject to any great challenge in this regard; we simply have no grounds for thinking that our current grasp of the mechanisms of selection is deficient in this way.

More importantly, there is an obvious disanalogy between the examples that Post employs, and, without the analogy, he cannot lean on the lessons from the history of science that motivate Papineau's model. The disanalogy is easy to see. Part of the reason it is rational to adopt the hypothetical equivalence in BL is that physicists *already have* a detailed grasp of the mechanisms of high-density plasma. Were this not the case, no right-minded physicist would offer BL as a serious hypothesis. No physicist would put forward a hypothesis that amounts to the following: "I hereby hypothesize that ball lightning is equivalent to high-density plasma—although I admit we have no idea what physical mechanisms constituting high-density plasma could possibly produce ball lightning and we currently have no grounds for thinking that our theory of the mechanisms of high-density plasma are in any way deficient." Yet Post is urging that we adopt a hypothetical equivalence regarding normative functions that is exactly analogous to this clearly unacceptable line of reasoning. He is proposing that we adopt an equivalence between noncausal, non-mechanistic norms of performance and descent from ancestral selective success, even though we have no idea how the mechanisms involved in selection could possibly create such norms and no grounds for thinking that our current account of those mechanisms is deficient. Contrary to Post's argument, the alleged equivalence in DFOR does not fit Papineau's model of ontological reduction.

Robert Brandon (1990) describes five categories of evidence that help us distinguish plausible from implausible adaptationist hypotheses, and Post appeals to these categories as potential empirical evidence for DFOR. Post's suggestion is that DFOR is open to empirical test by applying Brandon's categories. But we must wonder. The categories described by Brandon serve as evidence that a given trait is descended from selectively successful ancestors and, as I have said, evidence of descent from selective success is not sufficient evidence for the existence of noncausal, nonmechanistic norms of performance. Or, more modestly, evidence of descent from selective success might qualify as evidence for noncausal, nonmechanistic norms of performance if the equivalence in DFOR were already established on independent grounds. The evidence for one relata (descent from selective success) might qualify as evidence for the other (noncausal norms) if there existed an independent argument for the equivalence in DFOR. But no such argument has been given. So far as Post has shown, there exists not a shred of naturalistic evidence for the equivalence in DFOR.[11]

I conclude, therefore, that the theory of proper functions, because it flouts the directive in (EC), is an abject failure. None of the formulations of the theory specifies real-world mechanisms capable of producing noncausal norms of performance. Post's recent appeal to Papineau's model of reduction, though an interesting twist on the theory, does not solve the problem. There is, after all, an indefinite number of trial balloons that might be formulated regarding our concept of normative functions. Very few of those balloons, however, are worth floating. We need to have *some* naturalistic reason—some plausible evidence concerning the actual-world mechanisms involved—for taking any one balloon and letting it loose. No such reason has been given.

'Self-Perpetuation': A Genealogy of a Dubious Concept

I turn now to a further flaw in the theory of proper functions. Even if the worries raised above concerning the directive in (EC) could be alleviated, we must still address the question whether the theory adheres to the directive in (DD) against concepts dubious by descent. As we will see, it does not. Tracing just a portion of the genealogy of 'self-perpetuation' reveals that the concept of normative functions is deeply dubious by descent. That gives us additional grounds for rejecting the postulation of literal functional norms.

Dubious concepts run deep in the fabric of our worldview and color our orientation in ways we sometimes fail to perceive. In the case at hand, there is an equivocation in our concept 'self-perpetuation' that gives the theory of proper functions an illusory sense of plausibility. We discover the equivocation by tracing a single thread in the genealogy of the concept, namely, its explication in Kant's theory of natural purposes. Of course, the cultural ancestry of our concept 'self-perpetuation' extends far beyond Kant and is a messy tangle even within Kant's work. Still, getting our hands on a single historical strand is instructive—it makes us more self-conscious about the origins of our conceptual intuitions, increases the likelihood that we will recognize the theological vestiges that persist in our worldview, and forces us to adopt an orientation that cultivates the growth of an historical imagination. All this is true even if the genealogy offered is partial and even if it is only partly right.

I have already noted a point of agreement between Kant and advocates of selected functions. They agree that at the core of our concepts 'purpose' or 'normative function' is the further concept 'self-perpetuation.' For Kant, however, the self-perpetuating capacities of living things are the expressions of an elemental "formative power" that animates all life. It is worth rehearsing the passage that serves as epigraph to this chapter:

An organized being is then not a mere machine, for that has merely *moving* power, but it possesses in itself *formative* power of a self-propagating kind which it communicates to its materials though they have it not of themselves; it organizes them, in fact, and this cannot be explained by the mere mechanical faculty of motion. (Kant 1790, sec.65)

Clocks are mere machines composed of independent parts integrated to fulfill a specific function.

Though obviously purposive, their purpose is imposed from without. They are entirely passive. They do not incorporate external elements into their own form; they do not produce more of their own kind; nor, except through some external agency, do they maintain their internal integrity. Organisms, by contrast, accomplish all these tasks not from without but from within. They are essentially active. Nothing external to an animal or plant imposes or maintains its drive to reproduce, to grow, or to regenerate its parts in response to injury. All this is accomplished by powers emanating from within. The laws of Newtonian mechanics may explain the external purposiveness of a machine but lack the resources to account for the internal purposiveness of a living organism. Self-perpetuation in the realm of the living is underdetermined by Newtonian mechanics. We must, therefore, acknowledge the theoretical necessity of positing a formative power unlike any natural process or mechanism within the Newtonian repertoire. We are driven by the very nature of the phenomenon—the apparent purposiveness of living things—toward a form-giving power that appears to influence the domain of the mechanical but is not itself mechanical or causal. It is not, at any rate, something we can correctly conceptualize as a mechanism or a cause.

As I have said, advocates of proper functions may claim their advantage here. They claim to preserve as central the concept of self-perpetuation while discarding the appeal to formative powers beyond the ken of causal-mechanical explanation. This appears to be intellectual progress that illustrates and justifies the alleged fertility of concept location. But, as I have said, appearances deceive. It is simply not true that the theory of proper functions preserves the concept of self-perpetuation. To the contrary, there is a decisive shift in the meaning of 'self-perpetuation' from Kant's teleology to evolutionary theory. A little genealogy reveals that evolutionary theorists cannot coherently employ the concept as Kant did. The failure to keep these two senses separate accounts for the initial intuitive pull of the theory of proper functions.

To appreciate the equivocation, we must peek inside the artifice of Kantian teleology. In keeping with the critical project, Kant insists that the real nature of life is beyond the reach of human understanding. More precisely, he insists that we cannot know whether our understanding of life is correct. It may be correct, but the structure of our cognitive capacities is such that we must refrain from claiming knowledge in this case. The distinguishing features of all life, according to Kant, stultify human understanding. The nature of life is inscrutable for us. He says,

To speak strictly, then, the organization of nature has in it nothing analogous to any causality we know. Beauty in nature can be rightly described as an analogon of art because it is ascribed to objects only in reference to reflection upon their *external* aspect, and consequently only on account of the form of their external surface. But *internal natural perfection,* as it belongs to those things which are only possible as *natural purposes,* and are therefore called organized beings, is not analogous to any physical, i.e., natural, faculty known to us; nay even, regarding ourselves as, in the widest sense, belonging to nature, it is not even thinkable or explicable by means of any exactly fitting analogy to human art. (Kant 1790, sec.65)

Natural purposes are neither thinkable nor explicable in terms analogous to human art—not even thinkable! The recommendation that we conceptualize nature as we do artifacts, therefore, is not intended by Kant to illuminate the darkness. It does not resolve the fundamental unintelligibility. No amount of progress in biology can remove the inscrutability of life. We must settle for the heuristic—a regulative maxim—of conceptualizing living things the same way we conceptualize artifacts, namely, as end-directed. This provides us some traction in the study of life—it underwrites the discovery of some true, lawlike generalizations, though not (as we will see in a moment) the discovery of what Kant considered genuine laws. On Kant's view of life, we can never reach beneath the surface. (This is a feature of Kant's view that Ruse plays down. Had he considered the depth of Kant's pessimism about the possibility of biology as a genuine science, his enthusiasm for Kant's heuristic may have been tempered.)

Still, a protoscience of life is better than nothing. So how do we accomplish even that much? How, by conceptualizing living things as we do artifacts, do we discover true, lawlike generalizations about living organisms? The general steps are as follows. The study of life is possible for inquirers like us only if there are discoverable laws that apply to living things. Now, all genuine laws possess the robust universality and necessity that grace the laws of motion in Newton's physics. (This, just to be clear, is Kant's assumption, not mine.) The fundamental task then is to try to conceptualize living things in such a way that some suitable approximation to genuine laws is possible in biology. This is the crux of the matter—how to conceptualize living things so that we may discover as far as possible what is universal and necessary to all life?

Kant approaches this by drawing an analogy to Newtonian physics. Laws of physics are universal and necessary in so far as nothing happens by chance, in so far as everything that happens must happen. That nothing happens by chance is, for Kant, the "fundamental proposition of the universal science of nature" (section 66). By analogy, we can construct a proto-science of life if we can conceptualize the distinguishing features of living organisms, their capacity for self-perpetuation, as a specific instance of this "fundamental proposition." And, says Kant, we can conceptualize self-perpetuation this way. The claim is not that self-perpetuation is correctly conceptualized in terms of the fundamental proposition; the claim is that we that we can and, if we wish to be able to study living things at all, we must conceptualize it this way.

The fundamental proposition in the case of living things is not that nothing happens by chance but the more specific proposition that nothing in living organisms is in vain. Kant tries to encapsulate this proposition in what he calls the "principle of judging internal purposiveness in organized beings": "This principle, which is at the same time a definition, is as follows: *An organized product of nature is one in which every part is reciprocally purpose* [end—Trans.] *and means.* In it nothing is in vain, without purpose, or to be ascribed to a blind mechanism of nature" (Kant 1790, sec.66). The trick is to see parts of an organism as we would parts of artifacts, where the functional significance of each part is determined by a prior conception of the whole system and where each part contributes to the ends of all other parts and the ends of each part are sustained by the contributions of all the others. That, presumably, is how an intelligent mind—a supreme engineer, at any rate—would design an artifact. The system as a whole would be conceptualized in advance, and nothing in the system would be extraneous—each part would be necessary to the fulfillment of the ends of the whole. There would be no waste, no lack of direction, no lack of fit; everything would be positively charged with direction and purpose. And in so far as we conceptualize all of life this way—in so far as we impute this type of necessity to the entire realm of living organisms—the principle applies universally.

The final step in Kant's reasoning appears to be the most crucial. In order to conceptualize life in light of the principle of "internal purposiveness in organized beings," we must first conceptualize every living thing as animated by an intrinsic formative power of a self-propagating kind. We must see every token organism as animated by an intrinsic drive to perpetuate its specific form and as wielding an internal template that specifies and imposes its form. Despite its ultimate inscrutability, the postulation of this formative power is at the heart of Kant's theory of natural purposes. This is because it helps secure the proposition that nothing within a living whole is in vain. If all the parts of a living system are animated by this form-giving power, they will thus serve as both purposes and means to the ends of the whole. They serve as means by contributing causally to the functional ends of other parts and thus to the whole; they serve as purposes by executing functional tasks determined by the ends of the whole. This intrinsic power thus enables us to apply the principle for judging internal purposiveness to living systems, which in turn enables us to see the internal workings of living organisms as necessary and universal, which, finally, enables us to discover true, lawlike generalizations concerning the nature of life. The very possibility of a protoscience of biology rests upon the heuristic of Kant's inscrutable formative power.[12]

The centrality of Kant's formative power reveals the enormous theoretical stress on the *self* in his concept of self-perpetuation, and this is crucial for seeing how far Kant's use of the concept diverges from its role in evolutionary theory. For Kant, stress falls upon what exists intrinsically in each and every organism: every token is animated by a form-giving drive to perpetuate its own form, a form fixed by an antecedently existing archetype. This formative power resides within the individual organism and drives it toward the perpetuation of its species form. Hence, the notion of self-perpetuation in Kant is first and foremost a notion of *self* perpetuation. All the theoretical action—the drive toward perpetuating one's form—occurs within the individual living thing and serves as the fundamental center of command and control. Precisely this center of command makes possible the perpetuation of all living forms.

And it is from this abstract, theoretical vantage point that the contrast with evolutionary theory is starkest. While Kant puts the theoretical stress on an intrinsic form-giving power that operates internally, evolutionary theory does the very opposite. Evolutionary theory *distributes* the theoretical weight over a host of factors, including some that are irreducibly external to the organism. More importantly, evolutionary theory *decentralizes* the factors involved; nothing functions as a center of command and control because no such center exists. The very shaping of living things across generations depends largely on external and nonintrinsic factors in ways that Kant's view cannot abide.

Two considerations within evolutionary theory make clear the distributed and decentralized production of living forms. The first concerns the logic of selection explanations. Selection occurs when there is variation in heritable traits that result in differential reproduction. The mechanisms of selection include specific conditions of the local environment that give relevance to differences between organisms in their causal powers. These are external conditions that discriminate among varying organismic traits and that ultimately determine differences between organisms in reproductive success. Differences between organisms in their internal constitution are of course crucial to differences in selective success, but internal differences make a selective difference only with respect to external factors that discriminate between organisms in their powers to survive and reproduce. So the basic logic of selection explanations distributes theoretical weight broadly, placing great stress on external factors that discriminate among differing internal constitutions by virtue of differences in causal efficacy. This would be true even if there really existed some sort of Kantian formative power.

The second consideration concerns the holistic nature of development. The fundamental question concerning ontogenetic development is, What are the recurring resources responsible for the perpetuation of organismic forms? The answer for Kant, cast in the form of a regulative maxim, is a nonmechanical intrinsic drive—an inscrutable center of command and control—within each individual organism to perpetuate its species form. So the key elements of Kant's view include (1) an internal, intrinsic drive that is (2) beyond any known type of mechanism or causality, and that (3) strives to manifest and perpetuate its specific form. By contrast, contemporary theories of development reject the conjunction of elements 1 and 2 because they refuse the postulation of an intrinsic drive that is noncausal and nonmechanical. All contemporary theories agree that, in addition to the causal-mechanical structures within organisms, environmental factors are essential to explaining development. The theoretical action is clearly distributed across the organism and its relations to a host of external causal factors.

There is, however, the further question of decentralization, since recurring sources of development can be distributed without being decentralized. If, for example, some factors are more important than others—if some play a more central role in directing the processes involved in ontogenesis—then we may have grounds for retaining a centralized view of development. We may have grounds for retaining the conjunction of elements 1 and 3, or at least a thesis in the spirit of 1 and 3. We could, for example, postulate a drive or, if not a drive, an inherent tendency within genes to manifest and perpetuate the species form. Something like this appears to be the view in Richard Dawkins's 1976 gene-centered theory, according to which individual organisms are properly conceptualized as nothing more than the genes' (or the genotypes') way of making more copies of themselves. Genes serve as the theoretical center of command and control by orchestrating the development of the organism's body and behavioral repertoire in order to help ensure its own perpetuation. Individual organisms come and go much like the seasons, but genes, thanks to the relatively reliable copying processes through which they replicate, enjoy a relative form of immortality.

At the opposing end of the spectrum is the developmental systems theory. On this view, the recurring mechanisms of development are vast and, though the crucial effects of genes are certainly included, genes are afforded no particular authority or centralized power. Indeed, Eva Jablonka and Marion Lamb (2005), in a beautifully clear defense of this view, describe four distinct and substantive resources involved in the development and evolution of organismic forms, namely, genetic, epigenetic, behavioral, and symbolic. They also describe the different ways in which these four resources are organized and expressed and the ways in which each can affect and be affected by the others. The crucial general claim is that there appears to be no non-question-begging grounds for conceptualizing genes as centers of command and control. Genes are of undoubted importance in ontogenesis—they provide information[13] causally necessary for organismic development—but the same is true of causal factors in the other three resources. Development requires a host of epigenetic interactions, behaviors, and culturally transmitted symbols.[14]

It is sometimes difficult to know whether there is a genuine disagreement between these contemporary views of development and, if there is, exactly what the disagreement comes to. Whatever disagreements exist, however, they are of no consequence to my genealogical speculation. What matters is that, even if (just for the sake of argument) we were to adopt a gene-centered view of development, the recurring resources involved in development are nevertheless decentralized in at least the following sense. As indicated, all contemporary theories of development reject the conjunction of elements 1 and 2—we reject 2 because no one, not even the gene-centered theorist, appeals to developmental sources that are beyond the bounds of currently known causes or mechanisms, and we reject 1 because there is no sense in which genes contain an internal drive that is genuinely intrinsic in its operations. The latter point is of particular importance. The behavior of genes, as is well-known, is affected by an array of extrinsic factors—cellular, organismic, and environmental. We can change developmental outcomes by manipulating mechanisms internal to the gene, without doubt, but we can also alter development by manipulating mechanisms in the cell, elsewhere in the organism, even in the environment.

We are left then with element 3 and the alleged "natural striving" of genes towards self-perpetuation. Yet this too is compatible with a decentralized view of development so long as we are clear that the natural striving is not the effect of any power "intrinsic" to genes but rather an effect of genes in their normal cellular, organismic, and environmental contexts. Decentralization is clear from the fact that, absent the distributed factors identified by Jablonka and Lamb—absent the epigenetic or behavioral factors, for example—genes would never exhibit anything we might reasonably describe as a natural striving toward self-perpetuation. What strives to perpetuate itself, if anything does, is the *ensemble* of internal and external factors that contribute every generation to development.[15] And that is enough to show that nothing in the natural world, not even the causal effects of genes or genotypes, serves as a center of command and control for the development of living organisms.

To the contrary, the recurring production of life is the repeated effect of an ensemble of disparate elements, an ensemble with no conductor. These elements are transitory and vulnerable. They are vulnerable because subject to alteration or annihilation from many directions; they are transitory because the recurring causes that produce organisms include a host of factors that regularly destroy organisms. And, of course, the factors that produce and destroy living forms also contain sources of variation that sometimes serve as catalysts of evolutionary change.

In sum, when evolution occurs via selection, the mechanisms of perpetuation include those that comprise the selective regime imposed by extra-organismic factors. Further, even when nonselective factors are involved, the mechanisms of development comprise a decentralized ensemble of genetic, epigenetic, behavioral, and symbolic resources. The fundamental resources with which organisms perpetuate their forms are internal and external, scattered and transitory, and on a par with one another in terms of their causal efficacy. When, therefore, an evolutionary theorist asserts that selectively successful organisms perpetuate themselves, she is categorically *not* asserting that organisms are animated intrinsically to manifest and perpetuate their own form. The emphasis falls not on the 'self' in 'self-perpetuation' but on a centerless ensemble of causal factors. The contrast with Kant's formative power could hardly be starker.

This contrast, moreover, is unsettling. If contemporary theories of development are on track, then the forms of life that gardeners and breeders cultivate are not at all as Kant and his contemporaries thought of them. The coherence, stability, and persistence of living forms are, for Kant, an expression of something intrinsic to the individual organism. For us, by contrast, no one and no thing ensures the coherence or stability of living forms. No one is in charge; nothing comprises a center of command; no sentry protects against change or destruction. To the contrary, the human form and all the other forms we encounter are fleeting effects of recurring ensembles of scattered causal factors, and only the most obstinate failure of historical imagination could make us think otherwise. Where Kant would have us look for necessity in the relations among organismic parts, a necessity underwritten by an intrinsic drive toward wholeness, contemporary biological theory insists that all is contingent, that even the greatest wholes are the products of a precarious mix of causal factors, that all of life is fragile.

There is, finally, the brute fact that the theory of evolution by natural selection explains features of life that conflict with self-perpetuation. The theory explains the evolution of new species out of old ones as well as the process of extinction. The theory explains not only the perpetuation of certain forms but also their alteration and eventual elimination. Kant's formative power explains neither. Kant supposes that the distinguishing mark of living things is the capacity for self-perpetuation, but that is not the way evolutionary theorists regard the nature of life. Some forms of life are perpetuated briefly and then disappear; others persist but evolve into some other form, only to disappear later on; and all forms of life, even the most resilient, end in oblivion. Individual organisms, of course, pass through distinct developmental stages, and it is plausible that the final and deteriorating stages of life were positively selected for. Death, the cessation of the 'self,' is as integral to an evolutionary view of life as the periods of perpetuation that so impressed Kant. The concept 'self-perpetuation,' therefore, as it occurs in evolutionary theory, cannot mean what Kant meant by it.

This equivocation in 'self-perpetuation' is, I conclude, responsible for the prima facie (though mistaken) plausibility of the theory of proper functions. We inherit, via various cultural institutions, including the language of our ancestors, the Kantian construal of this concept; we then learn that the mechanisms of selection have the effect of perpetuating certain forms of life; and we conclude that the former (Kant's rendering of the concept) can be explicated and preserved in terms of the latter (the theory of evolution by natural selection). When seen from a distance, it all seems to fit so well; it all seems to make so much sense. And having come this far, it is tempting to conclude that, first, the mechanisms of selection do indeed underwrite the attribution of survival and reproduction as literal ends and, second, the attribution of literal norms of appraisal is well founded. This, I speculate, is the line of thought that seduces us toward the theory of proper functions.

Relinquishing the Concept Location Project

The intuitive appeal of the theory of proper functions rests upon a conflation regarding a concept at the intuitive heart of the theory—or so I have claimed. If that is right, then it is reasonable to generalize. Our concept of normative functions does not appear particularly parochial—nothing sets it apart from other concepts with a dubious history—and it is easy to see how gaps between old and new concepts are likely, especially when the old concepts are dubious by descent. We thus should conclude that the attempt to locate any dubious concept in a well-confirmed scientific theory is likely to invite the kind of equivocation just diagnosed. The claim here is historical not conceptual; there is no necessity that ancestral concepts should clash with new ones, merely the rational expectation of conflict. The expectation is rational in light of the fact that concepts dubious by descent are the product of our unchecked and culturally instituted biases, while the concepts and claims of our best-developed sciences have been, at least in some domains, subjected to the world.

The progressive naturalist should conclude that the concept location project, applied to dubious concepts, is a thoroughly conservative endeavor to be eschewed, not engaged. We may, in fact, conclude something quite specific, for the directive in (EC), combined with the directive in (DD), yields the following constraint on naturalistic inquiry:

Concept location project (CL): For any concept dubious by descent, expect that the concept location project will fail; expect, that is, that the dubious elements of the traditional concept will face revision or elimination as we analyze into and synthesize across the concepts and claims of all the relevant contemporary sciences.

The rational expectation expressed in (CL) has greater scope than either (EC) or (DD). It directs us not merely to withhold antecedent authority to this or that dubious concept but to turn our backs on an entire method of inquiry. It appears rational, that is, in light of our cultural history and with respect to concepts dubious by descent, to refuse altogether the task set by the concept location project.

And this means that a good deal of what currently passes as serious philosophical reflection must also be laid to rest. The aim of trying to save concepts of apparent importance by locating them in this or that scientific theory is a fruitless endeavor when the concepts are dubious by descent. As I say, the concept location project feels most attractive with respect to concepts most dubious—concepts inherited from our theological predecessors that appear to resist scientific understanding. Those are the concepts that strike us as most in need of preservation, especially if they concern the nature of being human, and yet precisely those are the concepts least likely to survive the assaults of progress in knowledge. That is why the methods of even our best naturalists are in need of reform.

There is, for example, something clearly self-defeating in Ruse's defense of the metaphor of design. Ruse is an engaging defender of evolutionary theory and a self-described evolutionary naturalist.[16] And yet, in trying to make sense of natural purposes—in endeavoring to preserve a concept of normative functions—he commits himself to a view that conflicts with an important implication of contemporary evolutionary theory. He insists that the practice of evolutionary theory requires the metaphor of a designing mind—the metaphor of a center of command and control—even though a radical implication of evolutionary theory is that the perpetuation of living forms is not under the control of a command center of any sort. Not even a metaphorical one.

Consider too the implications of the directive in (CL) for the theory of proper functions. In trying to make sense of natural purposes, advocates of proper functions claim that the mechanisms of natural selection, by sorting among competing forms of life and perpetuating those that best fit the demands of the environment, thereby produce a range of literal norms of performance. This is to conceptualize natural selection as a designer of functional norms—a mindless designer, to be sure, but a designer all the same—and it is precisely here that the theory of proper functions reverts to a center of command and control. Just look at the way the theory asks us to conceptualize the powers of natural selection. We are asked to conceptualize selection as a designer of functional norms, an allegedly natural process that contains the resources to produce standards of evaluation that survive the loss of underlying causal mechanisms. But this is every bit as unstable as Ruse's view: when advocates of the theory wish to boast of its naturalistic credentials, selection is described as causal all the way down. When, however, they wish to boast of the theory's capacity to preserve our concept of functional norms, selection is described as the source of norms that are decidedly noncausal. But advocates of proper functions have not yet explained how both boasts can be true. And for good reason![17]

The directive in (CL) also helps us better appreciate the plausibility of the alternative theory of functions that I endorse. Recall, first, the several objections facing the theory of proper functions: norms of performance do not analyze into evolutionary theory or synthesize across other theories; evolutionary theory does not justify the attribution of organismic ends, in which case the attribution of functional norms has no traction; and the intuitive power of proper functions derives from an equivocation in our concept 'self-perpetuation.' What remains intelligible after all of this is the attribution of functions minus the putative norms. What remains is the attribution of systemic capacities that contribute to specified high-level systemic capacities. This is to turn our backs on a dubious concept (and thereby refuse the project of concept location) in favor of a concept that coheres well with our best-developed knowledge of the world.

Conclusion

The proper attitude with respect to our conceptual categories is progressive, not conservative. Given what we know about ourselves—given, in particular, the tension between the theological ancestry of our concepts of human nature and the conceptual creativity required in our best sciences—it is rational to expect change in the categories with which we understand being human. And as the history of science attests, there has been enormous change in the categories in terms of which we understand the natural world quite generally. How could we fail to expect similarly radical changes in the way we understand ourselves—"children of nature" that we are?

The history of modern science has produced genuine growth in human knowledge. Not that there is anything intrinsically progressive about the quest for knowledge—things could have developed differently—but, as it happens, we now know more about the world we inhabit than at any other time in the history of human life on earth. We can now explain, predict, and control natural events in ways our ancestors could not. We also know from the history of modern science that such progress is more than mere accumulation.

It involves the creation of new conceptual categories to better conform our thoughts and feelings to the world, which in turn involves the destruction of ancestral categories. Certain things must be put behind us. What makes the concept location project so deeply objectionable is that, at its heart, it is the dumb refusal to put certain things behind us. It is the refusal to relinquish those concepts that seem to us, or seemed to our ancestors, to correctly specify in advance what we or the rest of the world are like. It is the default assumption of all theology and, alas, of a fair bit of contemporary philosophy.

SIX

A Persistent Mode of Understanding: The Psychological Power of 'Purpose'

We may no longer be thinking of a literal designer up there in the sky, but the mode of understanding persists. **MICHAEL RUSE, *DARWIN AND DESIGN***

Some concepts, including those involved in our traditional understanding of life and agency, are perpetuated in part by the conservative effects of culturally transmitted conceptual schemes. The directives discussed in previous chapters are intended to mitigate the retarding effects of concepts dubious by descent. It is implausible, however, to conclude that the cultural perpetuation of concepts is the whole story. It may not even be the most important part of the story. The cultural staying power of some of our concepts may be a more or less direct consequence of our psychological constitution. The impressive capacity for imitation among children and adults is plausibly one psychological capacity underwriting the generation and persistence of culturally instituted categories. Another likely cause of our conservatism regarding concepts is the fact, mentioned in chapter 3, that our orientation toward perceived threats is calibrated early in life and is resistant to change later in life.

There are, moreover, further capacities that plausibly underwrite the human inclination toward conceptual conservatism without the conserving effects of cultural evolution. At least some features of human psychology appear to be so deeply entrenched in the architecture of our minds that selective competition among culturally instituted concepts is probably too weak to drive them away. The focus of this chapter, at any rate, is a handful of capacities that appear to be entrenched in this way. As I illustrate regarding the specific concepts 'purpose' and 'function,' there is good prima facie evidence that the capacities that cause us to apply these concepts are not only entrenched but prone toward false positives. They incline us toward a form of conservatism regarding 'purpose' that leads us away from the truth about purposes. This is to illustrate the importance within a progressive orientation of the directive to withhold antecedent authority to concepts dubious by psychological role.

I begin with the suggestion that our conscious attributions of 'purpose' are plausibly the product of an entrenched system of mind, namely, the theory of mind system. The thought is that the capacities that cause us to apply 'purpose' are the same capacities that cause us to conceptualize certain objects as minded agents. The latter capacity, of course, generates an abundance of false positives. We readily see—we cannot help but see—minded agents or telltale effects of minded agents at nearly every turn, even when none is present. If this line of thought is plausible, then it is reasonable to expect that the capacities causing us to conceptualize certain objects as purposive are similarly prone toward false positives. This, I suggest, is evidence enough that the directive concerning concepts dubious by psychological role—the directive in (DP)—applies to Ruse's metaphor and the theory of proper functions.

The power of (DP), moreover, is enhanced considerably in this case when combined with the directive concerning our evolutionary history. Or so I wish to argue. My strategy is to argue by analogy. I begin with the directive in (A)—a plausible corollary of the directive in (EH)—that every cognitive or affective capacity of the mind is endowed with one or more systemic functions that are anticipatory. The suggestion is that all our capacities, including those that do not intuitively strike us as future oriented, nevertheless serve to orient us toward future elements of our internal or external environment. And, indeed, as recent theorists have analyzed into our capacities for vision, memory, and dreaming, they have confirmed the plausibility of attributing systemic functions to low-level systemic mechanisms that are anticipatory in nature. And, in so doing, these theorists have discovered that our initial understanding of these capacities is often wide of the mark. As we analyze inward, the concepts in terms of which we conceptualize the high-level capacities are changed, and sometimes the changes are substantial.

What I want to suggest then is that we approach our capacity to apply the concept 'purpose' in analogous terms. The suggestion is that our apparently natural inclination to conceptualize parts of living organisms in terms of natural purposes is itself caused by low-level mechanisms endowed with systemic functions that are anticipatory. So the question is, What are the anticipatory, systemic functions served by the mechanisms that incline us to see the parts of living organisms as purposive? The correct answer, at present, is that we do not know. We can, however, draw the analogy to its full conclusion by suggesting that, whatever the correct answer, our understanding of the inclination to attribute such purposes is probably in error. As with vision, memory, and dreaming, it is plausible that the fruits of analyzing inward will be the discovery of anticipatory, systemic functions that alter the way we conceptualize our aesthetic responses to living things. The importance of combining the directives in (EC), (EH), and (A) should not be underestimated.

The final section of this chapter rounds out the discussion of part 2 by applying all of my directives at more or less the same time to Ruse's metaphor and to the theory of proper functions. It is helpful to see that the efficacy of any one directive is often enhanced by the application of others. It is also crucial to see that our traditional concept of normative functions has, in light of these directives, lost whatever theoretical life it once had. We now know too much to seriously consider investing the resources trying to save it; 'normative function' has become a relic of a worldview that progress in knowledge has already put behind us. It is only the conceptually conservative orientation of contemporary philosophy and the concomitant failure to cultivate a progressive orientation that make us think or feel that this concept retains any value.

On my interpretation of Darwin's rhetoric in the *Origin,* the capacity that inclines us to attribute 'purpose' to the parts of living things is among the dispositions of mind that produce in us a failure of psychological imagination. The failure results from the fact that we cannot observe from the first-person perspective the anticipatory, systemic functions served by the felt inclination to apply 'purpose.' What is available to conscious awareness is an affective state that we regularly misinterpret as evidence that the parts of living things are indeed endowed with natural purposes. This is the central theme of the present chapter. But it is worth noting that the lesson of this chapter also reveals an associated failure of historical imagination: the failure to sum up in our minds myriad evolutionary pressures for capacities that were both anticipatory and beyond the reach of conscious awareness. Nonconscious capacities that were reliably and efficiently anticipatory would have been of much greater value than any capacity for conscious awareness of the operations or contents of one's mind. It is a plausible evolutionary hypothesis, at any rate, that the anticipatory systemic functions of our nonconscious capacities take us much closer to the truth about our selves than the meager offerings of conscious awareness. And this affects our methods of inquiry. In adopting a progressive orientation, we frame our inquiry in terms of anticipatory functions, we analyze inward from there, and we discover a plausible explanation of our tendency to see living things as endowed with purposes without the slightest reference to values or norms of performance.

'Purpose' and the Theory of Mind Theory

The second premise of Ruse's argument for the metaphor of design asserts that we are inclined by our psychological constitution to conceptualize living things the same way we conceptualize the products of human art. The claim, in effect, is that the capacities of mind that apply the concepts 'purpose' and 'function' are promiscuous, seduced by features of living things that bear some analogy to the features of artifacts. The very architecture of our minds, it seems, ensures that we cannot help but see and respond to living things as if they were the products of a mindful process of creation. This is Ruse's second premise.

It is odd that Ruse, an avowed Darwinian naturalist, fails to offer available evidence for this premise. The psychologists Deborah Kelemen and Frank Keil, for example, describe experiments that appear to support the claim that we are, beginning around five years of age, inclined by the structure of our minds to conceptualize all manner of objects in terms of purposes or functions. Young children, in Kelemen's (2004) phrase, even show signs of being "promiscuous theists"—or at least promiscuously inclined toward concepts of purposiveness that seem to entail some sort of agency.

Although the views of Kelemen (1999) and Keil (1994) overlap, there are important differences. According to Keil, the best interpretation of the experiments conducted so far is that as the mind of a young child develops, it acquires a more or less autonomous and domain-specific capacity that is sensitive to certain features of living things. This capacity, according to Keil, is dedicated to the application of various concepts, including 'function' or 'purpose,' in the presence of living things. Kelemen argues that the experiments so far better support a different claim. On her view, what emerges in the mind of young children is not an autonomous sensitivity to features of living things but rather a sensitivity parasitic on a more entrenched psychological capacity. Kelemen begins with the prevalent assumption that most children by the age of four have acquired a theory of mind capacity. But she further argues that in the course of the following year, the child's capacity for conceptualizing other minds is generously extended to all manner of animate and inanimate objects, causing them to conceptualize the components of those things as purposive. This is an intriguing story, so I begin with it.

The *theory of mind theory* is not specifically about the concepts 'purpose' or 'function.' Instead, as its name suggests, it is about concepts associated with capacities of the mind, including 'intention,' 'belief,' 'desire,' and more. In particular, the theory addresses the evident capacity of human beings to recognize and respond to objects that appear to be minded. This capacity appears to emerge at a startlingly early age, causing us to conceptualize apparently minded objects in terms of 'intention,' 'desire,' and the rest. Several considerations converge on the theory of mind theory. I will survey just three: the general framework and two sets of provocative studies that support the theory. With these considerations on the table, I then return to the concepts of 'purpose' and 'function.'

Nicholas Humphrey (1986) tells of the three months he spent living in Rwanda, observing the daily activities of a group of families of gorillas. After several weeks of observation, Humphrey grew perplexed by the apparent banality of gorilla daily life. Upon waking, the gorillas feed and then set out to other known sources of food or search for new ones. After additional feasting and an afternoon siesta, they eat once more, fashion their beds for the night, and finally settle in. That is a typical day in the life of these families. Overt conflicts are infrequent. There are squabbles over who grooms whom, who eats what, who sleeps where, and the like, but almost all are resolved peacefully. The gorillas do, of course, fight among themselves, and some conflicts become deadly, especially those concerning access to mates. But violent conflicts are rare; stability and calm are the norm. Humphrey wondered why this is so.

He speculated that the absence of overt social turmoil is a consequence of the fact that gorillas are endowed with a high degree of social intelligence. Conflicts occur on a daily basis, since questions of food and grooming are far from trivial, but these problems do not escalate because they are quickly resolved. And their quick resolution is the happy effect of the affective and cognitive capacities involved. Humphrey's suggestion is that the gorillas he observed are naturally equipped to identify and respond to the mental states of their family members and social peers. They have the capacity to interpret the behavior of other gorillas as out-ward expressions of internal cognitive and affective states, as well as the capacity to respond by behaving in ways that are similarly interpreted by others. This is to assert that Rwandan gorillas are neurologically and psychologically endowed with the capacity to conceptualize the movements and sounds of certain objects—other gorillas, at least—in terms of a host of mental concepts or, more modestly, in terms of analogs of concepts that we categorize as mental, namely, 'intention,' 'belief,' and so on. What better explanation for the observed banality of gorilla daily life?

The core of Humphrey's hypothesis is echoed in recent literature, and differences among primate species in their capacities for social intelligence have been vigorously debated. Richard Byrne and Andrew Whiten, in two collections of essays (Byrne and Whiten 1988; Whiten and Byrne 1997), focus on what they term the *Machiavellian intelligence* of the primate mind. The capacity to identify and respond to other agents in terms of relative social status is, they suggest, among the most central constitutive capacities of the primate mind. This includes the capacity to recognize one's own social standing, the capacity to recognize differences in social rank, to form and maintain social alliances, to recognize and respond to potential violators, and more. More recent works, including one by Michael Tomasello and Josep Call (1997), provide excellent syntheses of the proliferation of work in these areas. Tomasello (1999) and, more recently, Dorothy Cheney and Robert Seyfarth (2007) defend Humphrey's central thesis of social intelligence, while also bringing into focus the crucial differences between humans and other primates with respect to their theory of mind capacities.

Although it is easy to speculate on how the capacities involved in social intelligence emerged in the course of primate evolution, compelling evidence is only just beginning to emerge. Nonetheless, we—along with Cheney, Seyfarth, Tomasello, and others—should wonder exactly how the capacity for primate social intelligence is implemented in us. *Homo sapiens* has lived in social units for all of recorded history and for vast spans of unrecorded history, as suggested by archeological remains. If we, like Humphrey's gorillas, are natural-born psychologists, how—by virtue of what capacities—do we do it?

The nice thing about this question is that we already have the beginnings of an answer that rests not on evolutionary speculations but rather on the fruits of analyzing inward and synthesizing laterally. This brings me to the first set of provocative studies concerning the theory of mind theory. Some of the most compelling evidence comes from studies employing the false-belief test, a simple experiment designed to determine whether subjects have the capacity to attribute beliefs—false beliefs, in particular—to other individuals. Experimental subjects are presented with some variation on the following theme. Imagine a scenario in which a beautiful young girl, Cassie, is holding a chocolate candy. She carefully places her chocolate in a basket for safekeeping and then leaves the room. A short while later Annie, another lovely girl, comes in the room, noses around, and discovers the chocolate. She removes it from the basket, places it in a box on the other side of the room, and departs. Having heard this much of the story, subjects are then asked two control questions followed by a single test question:

- In the beginning, where did Cassie put the chocolate?
- Where is the chocolate now?
- When Cassie comes back, where will she look for her chocolate?

Several interesting discoveries have resulted from the widespread use of this test. The first is that most three-year-old children fail the test; they say that Cassie, when she returns to the room, will look for the chocolate in the box where Annie put it. The second discovery is that most four-year-olds pass the test. During the fourth year of development, most children acquire a capacity that three-year-olds lack, namely, the capacity to attribute a mental state to another person that is clearly distinct from the child's own mental state. The third discovery is that most ten-year-old children with Down syndrome pass the test; they rightly say that Cassie will look for the chocolate in the basket, not in the box. The fourth discovery is that most twelve-year-old autistic children—fully 85 percent of them—*fail* the test. This last finding is striking in light of the general lack of social intelligence exhibited by autistic children.

What explains the fact that three-year-old normal children and twelve-year-old autistic children fail the test? Apparently they do not have the conceptual resources with which to distinguish the beliefs of another person from their own beliefs. In three-year-olds, the resources have yet to develop; in autistics, they have yet to develop or are defective in some way. What explains the fact that four-year-old normal children and even ten-year-old children with Down syndrome pass the test? They apparently have the conceptual resources that three-year-olds and autistics do not. In particular, they can attribute to another person a belief that is distinct from their own beliefs, indeed, a belief that they (the subjects) know to be false.

The intriguing theoretical question concerns the correct characterization of this difference in cognitive resources. In an excellent discussion of this topic, Alan Leslie (2000) ventures the hypothesis that four-year-olds have, and three-year-olds lack, a matured mechanism of selective attention. More fully, the architecture of the human mind from an early age includes a mechanism that causes us to focus attention upon specific sorts of external stimuli. Several experiments even seem to suggest that a developmental precursor to this mechanism is up and running during infancy.[1] What the false-belief experiments suggest is that around the age of four this mechanism has matured and is now triggered by the perception of stimuli that are characteristic of minded agents. These stimuli trigger a mechanism that focuses attention on the relevant object, namely, an agent, and these stimuli also trigger a set of representations or knowledge structures that cause us to conceptualize the object as possessed of and moved by internal mental states. Leslie calls this mechanism the *theory of mind mechanism;* the mechanism and the associated set of representations are collectively called the *theory of mind theory.*

(The name "theory of mind" may invite needless confusion. Leslie is not claiming that four-year-olds are little scientists, armed with a theory of the human mind, wondering whether or how it will be confirmed. He is claiming that the architecture of the human mind includes a developmentally entrenched attention mechanism and a set of dedicated concepts especially prone to fire in a specific domain, namely, when stimuli characteristic of minded agents are perceived. Some theorists, in order to avoid this confusion, refer to the attention mechanism and associated concepts collectively as the capacity for *mind reading*—for identifying and interpreting the workings of other minds.)

An operative theory of mind mechanism appears to be the very thing that autistic children lack. Autism is diagnosed behaviorally and is neurologically based. Typical indicators include impaired social skills, language delay, and the absence of pretend play. At least part of the deficit includes the inability to conceptualize other agents as agents. The following description of an autistic child makes this point vividly: "On a crowded beach he would walk straight toward his goal irrespective of whether this involved walking over newspapers, hands, feet, or torsos, much to the discomfiture of their owners. His mother was careful to point out that he did not intentionally deviate from his course in order to walk on others, but neither did he make the slightest attempt to avoid them. It was as if he did not distinguish people from things, or at least did not concern himself about the distinction" (Kanner 1943, 232). If, around the age of four, a relatively mature theory of mind mechanism comes online in normal children, then it is reasonable to speculate that precisely this is what fails to occur in the development of autistic children.

It is also reasonable to wonder, however, whether the deficits associated with autism stem not from a broken or missing theory of mind capacity but rather from the more global deficits associated with mental retardation, including, for example, poor memory, poor problem-solving ability, and so on. Reasonable as it is, the evidence so far is against it. As Leslie notes, 25 percent of autistic children are not mentally retarded but still suffer impairment in social skills, language delay, and a failure to engage in pretend play.[2] Moreover, Leslie and Thaiss (1992) adapted an innovation on the false-belief test developed by Zaitchik (1990), and the results tell heavily against the suggestion that autistic deficits are the result of more general impairments. Leslie and Thaiss began by tutoring subjects on the workings of a Polaroid camera. After they were familiar with the camera and the representational nature of the photos it produces, subjects are asked to take a picture of a toy cat sitting on a chair. The photo is removed from the camera and placed facedown on a table, and the toy cat is moved from the chair to the bed. Subjects are then asked three questions analogous to those in the false-belief test:

- When you took the photo, where was the cat?
- Where is the cat now?
- In the photo, where is the cat?

Leslie calls this the *out-of-date photo test.* Two results stand out. The first is that while most normal three-years-olds failed and most normal fouryear-olds passed, some of the four-year-olds who failed the photo test had earlier passed the false-belief test. That is, for normal children, the false-belief test appears slightly easier than a structurally analogous test that involves the representational content of photos. This is at least some (very weak) evidence that four-year-olds are better attuned to facts about minds than to facts about photographs.

The second result concerns the performance of autistic children. Nearly all the autistic children who earlier failed the false-belief test passed the out-of-date photo test! Twelve-year-old children with autism who were unable to distinguish the beliefs of another person from their own beliefs nevertheless succeeded in distinguishing the out-of-date content of a photo from the actual state of affairs. Now, distinguishing the representational content of the photo from the actual state of affairs clearly seems to involve conceptual and computational resources akin to those required to distinguish the beliefs of another agent from one's own beliefs. There is, at the very least, a comparable degree of conceptual sophistication. So what appears different is not the level of requisite general mental ability but rather the specific nature of the affinities and concepts involved. Children with autism appear to suffer from a deficit keyed quite specifically to the perception and conceptualization of other minds.

Interestingly enough, the second set of studies that support the theory of mind theory rests less on the false-belief test and more on a range of evidence that human infants as young as nine months are attuned in important ways to the minds of other persons. The theory sketched in Tomasello 1995 and 1999 is instructive.[3] Tomasello (1999) reviews studies suggesting that between the ages of nine and fifteen months, human infants begin to engage in a variety of joint attentional behaviors. As early as nine months they interact with adults by following their gaze, jointly focusing on a shared object, imitating the behaviors of adults, and so on. This suggests that infants are beginning to tune in to the minds of other persons by the age of nine months. By thirteen months of age, moreover, infants endeavor to attract the attention of adults to objects. They do this by holding an object up or pointing at it.[4] As Tomasello points out, the emergence of these and other joint attentional behaviors calls out for explanation. His suggestion is that infants around this age are beginning to understand other persons as *intentional agents.* Something is an intentional agent if it has goals and pays attention to objects in its environment in an effort to satisfy its goals. The suggestion is that very young infants come to conceptualize and perceive other persons as motivated toward the satisfaction of their own goals. This is not to attribute to the infant the full-fledged mental concepts of the theory of mind theory; talk of goals and attention should not be conflated with talk of mental states such as intentions, desires, and beliefs. It is, however, to attribute to the infant the developmental rudiments from which mental states may later emerge.

And that is indeed how Tomasello conceives the matter. Somewhere around four years of age, when most young children begin to pass the false-belief test, children gradually cease being mere intentional agents and grow into what Tomasello calls *mental agents.* At this stage, children understand others not simply as endowed with powers of attention that they use in pursuit of goals. They begin to see others not merely as purposive but as purposive by means of more or less identifiable mental states, by means of thoughts and desires that may (but may not) be expressed behaviorally. And one distinctive feature of Tomasello's view is that becoming a mental agent not only explains the emergence of our social intelligence but is also partially explained by the social context in which the mental agent develops. The demands of social life contribute to the emergence of the developing child's theory of mind: "the transition to an understanding of mental agents derives mainly from the child's use of intentional understanding in discourse with other persons in which there is a continuous need to simulate other persons' perspectives on things, which often differ from the child's" (Tomasello 1999, 179).

We have then the following three considerations: the larger framework suggested by Humphrey, evidence for an operative theory of mind in four-year-olds (based on false-belief tests), and evidence for emerging elements of a theory of mind during infancy (based on joint attention tests). When taken together, these considerations suggest that some story more or less along the following lines must be true: the reason we are so attuned to other minds is that we are natural-born detectors of social agents—of objects, at any rate, that exemplify features characteristic of social agents. We are cognitively and affectively outfitted very early in development to preferentially attend to certain sorts of stimuli and, in response to those stimuli, preferentially conceptualize the relevant objects in terms of mental properties. We are oriented to conceptualize certain objects in terms of internal mental states central to agency, including states such as desires or intentions. This is the theoretical heart of the theory of mind theory.

Against the backdrop of this theory, Keleman hypothesizes that, after the theory of mind capacity matures by four years of age, the relevant capacities are then extended to other concepts, 'purpose' and 'function' in particular. The extension of these capacities appears to occur by five years of age. Keleman reports several experiments that seem to show children around the age of five and up to the age of eleven or twelve promiscuously applying the concepts 'function' and 'purpose' to animate objects such as horses and to inanimate objects such as rocks. And it is interesting that religious background is not a relevant factor. As Kelemen (2004) reports, studies by Evans (2001) reveal that even the progeny of robust atheists are reckless at this age in their use of 'purpose' and 'function.' However, as Kelemen further describes, this apparently natural disposition to see the world in terms of purposes begins to moderate around the age of twelve. Most twelve-year-olds begin to adopt the culturally pervasive view that inanimate natural objects are not purposive, and those reared in homes with little or no religiosity begin to withhold function attributions to the parts of living things as well.

Kelemen's speculative hypothesis regarding these developmental patterns is interesting. By four years of age, children conceptualize other persons in terms of certain mental categories; they naturally see other persons as moved by internal desires and directed by internal intentions. This is to see others as purposive in this sense: the behavior of other persons is conceptualized as aiming at some intended goal or end. What happens as four year olds mature further is something like the following. They see other persons—agents that they naturally conceptualize as possessed of various ends—employing artifacts in pursuit of their ends, and this causes children to conceptualize the artifacts themselves as being for the ends of intentional agents. They come to see artifacts, that is, as endowed with functions relative to the mental goals of the agents involved. Then, in the crucial step that leads to promiscuity, children extend this entire mode of understanding indiscriminately to living things and to inanimate natural objects. That is, as they mature throughout their fifth year, already conceptualizing artifacts as endowed with purposes, children then extend the same mode of understanding to all manner of objects. They do this, Kelemen suggests, because their experiences force upon them the need to conceptualize living organisms and inanimate objects in some way or other. And since they already have an operative understanding of artifacts (derived from their entrenched understanding of agents) ready to hand, they find it easy and perhaps natural to conceptualize livings things and even nonliving things in the same way they conceptualize artifacts.

This is to see the tendency in young children to conceptualize living things as artifacts as the default position. If Kelemen is right, the disposition to see all manner of objects as purposive or functional is a natural and unfettered effect of the structures of our minds. And it is only through learning, through exposure to beliefs concerning the nonpurposiveness of rocks or clouds, that children learn to check this natural disposition. What strikes young children as most natural, and what adults have to be talked out of, is a quite specific instance of a tendency to see the world in ways that it is not. This illustrates vividly the kind of disposition of mind that, if real, generates an abundance of false positives and, as a result, failures of psychological imagination.

Although Keil's view overlaps Kelemen's in several ways, he rejects the claim that concepts of purposiveness develop out of concepts of mind. He rejects the hypothesis that the capacities that constitute our theory of mind are the same ones that cause us to apply 'purpose' or 'function,' on the grounds that experiments seem to show five-year-olds discriminating between the application of certain mental concepts (such as 'intention') and the application of certain purposive concepts (such as 'function'). It is interesting, for example, that although five-year-olds seem to prefer descriptions of diseases in functional rather than merely mechanical terms—they find it natural to think of the disease-causing agent as directed toward a goal—they were unwilling to attribute any knowledge to the disease-causing agent. The disease-causing agent was seen as goal directed but not as knowing what it was doing. This, on the face of it, might suggest that the conceptualizing capacities of five-year-olds discriminate between the application of mental concepts and purposive concepts.

As with any novel, interesting research program, there is room to question the experiments and hypotheses offered by Kelemen and by Keil.[5] Such doubts, however, are not to the point. Whether we side with Kelemen or with Keil, or whether we try to develop a third option, we can, I think, draw the following modest conclusions. First, there exists a small but growing body of literature dedicated to assessing the question whether concepts such as 'purpose' are entrenched in our psychology. Although the literature directly concerned with 'purpose' is relatively small, the literature concerning the theory of mind theory is much larger and more robust. And, as suggested by Kelemen's hypothesis, the capacity to attribute an intention to another object seems to carry with it the capacity to conceptualize that object as having an end or goal, which in turn seems to include the capacity to conceptualize the agent's actions as functional, as fulfilling the agent's goals. And since the capacities posited within theory of mind are thought to run deep in our psychology, the capacity that causes us to apply 'purpose' probably does the same. This literature bears directly on the thesis of Ruse's second premise, and the verdict at this point, though by no means decisive, appears favorable.

The second conclusion is that since the capacities posited in the theory of mind are known to generate false positives, we ought to suspect the same of capacities that cause us to conceptualize the parts of plants and animals as purposive. This seems a reasonable presumption. We know, after all, the unavoidable inclination to see mindless objects as minded or inanimate things as animated. It is surely reasonable to expect that our felt sense that some things are purposive is a similar sort of inclination, causing us to see and feel as purposive things that have no purposes.

I do not take myself to have established the truth of either claim. Nor do I think that anyone else has developed a sufficiently convincing case. There is, however, reasonable presumption—reasonable because grounded in relevant experimental findings—that 'purpose' is indeed dubious by psychological role, and that is all I need in order to insist that it is rational to require of our methods of inquiry that they include the directive in (EH) concerning our evolutionary history and the directive in (DP) concerning concepts dubious by psychological role. The presumption that 'purpose' is apt by virtue of our constitution to lead us astray is enough to show that we ought to divest it of whatever antecedent authority our psychology tends to grant it.

Before concluding this section, I pause to highlight the sorts of conflicts we suffer as a consequence of our constitution. Autistic children tend not to conceptualize objects in their environments in mental terms; they tend not to conceptualize persons as things with minds. I take it for granted, moreover, that they tend not to notice this about themselves; they do not suffer the fact that a piece of their cognitive architecture is missing or broken. Similarly, persons without autism, though deeply disposed to conceptualize certain objects as mindful agents, tend not to notice this disposition in themselves. Seeing or experiencing the world in agential terms is embedded in our basic orientation toward the world, but rarely do we notice its effects. It is only when we meet someone who is missing the requisite cognitive capacity that we acquire any feel for the extraordinary capacity we have. We nonautistics are, to extrapolate from Ross and Ward 1996, naive realists concerning the existence of other minds. And, if I am right, we are similarly naive concerning the reality of norms of performance. We tend to feel and think that the view generated in our cognitive and affective field of vision, especially our depiction of minded agents and purposive traits, is an unfiltered and unbiased representation. Our capacity to consciously feel and see reality in these ways is simply a given; nothing in what we consciously feel intimates the extent of the distortion and falsification involved.

The cure for our naive realism, if there is one, is to recognize that we have little or no first-person access to the actual workings of our cognitive and affective capacities and that the way things appear to us in conscious awareness is often mistaken, and pervasively so.[6] The cure also requires that we develop strategies for inquiry that diminish or eradicate the retarding effects that naive realism typically exerts. The directives in (EH) and (DP) are intended to help accomplish that, and this is to say that we must bring to bear other, quite distinct psychological capacities in an effort to correct the falsehoods foisted upon us by our conscious capacities.

Evolutionary History and Anticipating the Future

The directive in (EH) is motivated partly by the assumption that a framework informed by facts of our evolutionary history helps us distinguish psychological structure from mere noise. If so, then correcting a failure in our historical imagination helps correct failures in our psychological imagination. It does so by directing our attempts to analyze inward toward the actual systemic functions possessed by low-level mechanisms. This is the crucial connection between (EH) and (DP) and also the connection between these directives and the expectation of conceptual change in (EC). The aim of this section then is to illustrate the power of these connections with respect to the mechanisms that cause us to apply the concept 'purpose.'

I begin by considering these directives in studies of human vision, memory, and dreaming. As we will see, the results of those studies have altered our prior understanding of the high-level capacities involved. What we used to think about the function of dreaming, for example, is hard to sustain in the face of recent discoveries. My hypothesis is that the same fate awaits our capacity to conceptualize the parts of plants and animals in terms of purposes. Once we frame our inquiry in evolutionary terms and inquire into the relevant systemic functions, our prior confidence that the parts of plants and animals must indeed be possessed of functional norms is diminished.

Recall the directive in (A), namely, that for any psychological capacity of any minded organism, expect that among its most prominent systemic functions is the anticipation of some feature of the organism's environment. Now, some capacities are anticipatory not because of ancestral evolutionary pressures but simply because of the constraints imposed by the materials out of which we are constituted. To get the job done, all our capacities must anticipate the future by operating within the bounds of chemical and physical laws. Consider in this regard our capacity to visually perceive the correct location of a fast-moving object. Imagine that you are a right fielder especially adept at catching fly balls. Unless you are also a student of recent theories of vision, you probably think that the visual image of the moving ball produced in your mind is a real-time depiction of the actual location of the ball. But it is not. The information about the location of the moving ball, by the time it is routed from your retina and then processed in your occipital lobe, is already out of date. Neural processing, no matter how swift, takes time; the chemical and mechanical processes that constitute our visual system operate within the bounds of the physical. Your brain, however, compensates. It does a quick bit of calculating and produces an image of the ball's location not where it was last seen but rather where it most likely is at this, or perhaps at the very next, instant. This is what explains your adeptness in right field—your visual system, it turns out, is an anticipatory marvel. It produces representations not of how the world was when you perceived it but rather how the world will be in the future, when your actions depend crucially on how things are at that time.[7]

We have here a capacity that makes us feel that our visual experience of a moving object is keyed to reality, to the immediate "now," though the feeling is an illusion wrought by the nonconscious, low-level workings of our visual system. To know this is to have a lovely bit of knowledge concerning the workings of our seeing capacities. And this bit of knowledge—the fruit of analyzing into the visual system—forces upon us a corresponding bit of reconceptualization concerning our capacity for sight. No longer can we conceive of the visual system as a passive device for depicting things in real time, at the instant of seeing. Our system "depicts" the location of moving objects as they will likely be in the very near future and not as they really were at the instant of seeing. Naive realism concerning our vision of moving objects is out of the question.

Now consider memory. The human capacity for memory comprises several more or less distinct memory systems, including what Martin Conway terms *autobiographical memory.* On Conway's view, autobiographical memory is composed of two general components: (1) a store of encoded information concerning one's sense of oneself and (2) a working memory (Conway calls it the *self-memory system)* dedicated to one's sense of oneself. Component 1 is a product of one's history; it contains memories that inform us about the particulars of our selves, about what we are like as unique individuals. Component 2 is an online workspace in which one's current interactions with the world are represented and processed. It is significant that the processing in working memory is constrained by more or less structural elements of oneself, including one's personality, attachment strategies, perceived discrepancies between motives and accomplishments, and the like.

These two components together serve the function of updating or revising one's memories of one's self, and if Conway's view is correct, this systemic function is fulfilled mainly with respect to the goals or motives imposed by standing features of one's personality, attachment strategies, and so on. This latter point is particularly intriguing. The store of past information, as I say, contains memories that bear on the sort of person one is, but these memories are retained and revised by virtue of their relevance to the goals or motives that partly constitute one's personality. Consider, for example, the difference between personalities oriented toward achievement, where the salient goals include mastery of skills, independence from others, power over others, and so on, as opposed to personalities oriented toward intimacy, where the salient goals include close communication, interdependence, and so on. Conway's hypothesis is that the creation and maintenance of autobiographical memories will differ dramatically between persons whose personalities differ in these ways, precisely because the difference in goals determines which experiences are saved or recast in memory.

A host of experiments appear to confirm Conway's hypothesis. There is compelling evidence, for example, that the sorts of emotional memories brought into the self-memory system tend to track differences in personality-based goals. For persons geared toward intimacy, the affective experiences most vividly remembered and readily retrieved are those concerning significant others, typically emotions associated with acts of friendship and love. For persons geared toward achievement, by contrast, the affective experiences remembered are those concerning the achievement of (or the failure to achieve) mastery or independence. What is most readily retrieved is the glow of success or the humiliation of failure. The very construction of autobiographical memories by the working self appears keyed to the motives or goals that are constitutive of one's personality type.[8]

It is perhaps no great surprise that we tend to remember emotionally charged events that fall readily under the purview of our standing motives, for those are the things that, given the ingredients of our personality, we care for most deeply. But what is surprising is the extent to which the memory system involved is oriented toward the agent's future. If the memories stored are those that best fit our personality-based goals, then this serves as a source of information about the objects and persons in our environment relevant to the future fulfillment of our goals. Our memories are concerned with the past effects of objects and persons in our environment because of their relevance to the future satisfaction of our personality-based goals. Being able to recall which objects and persons were helpful in the recent past and which were harmful is a potent resource for the future satisfaction of those goals.

Here we have a more complex illustration of the way in which discoveries of nonconscious processes force us to reconceptualize the conscious capacity we formerly thought we understood. Memories are naively assumed to be cumulative and more or less accurate records of reality, but on Conway's theory this is not the case. The problem is not the usual sorts of decay or distortion that afflict any memory system but a straightforward byproduct of the anticipatory, systemic function that the memory system serves. Autobiographical memory appears tailored to the future efficacy of one's standing goals or motives. Its function is to record and revise remembered experiences insofar as doing so increases the probability that our personality-based goals are satisfied in the near future. As in the case of seeing moving objects, our initial understanding of our own capacity is altered by analyzing inward and discovering the anticipatory, systemic functions of the nonconscious processes involved.

Consider, finally, the nature of dreaming, for that too is deeply anticipatory. Indeed, the nonconscious processes involved in dreams are so strange and surprising that a major reconceptualization is already under way among scientists who study sleep and dreams. Consider just one way in which dreams are anticipatory, namely, the relation between the especially florid dreams during rapid eye movement (REM) sleep and our emotional orientation toward the world while awake. Our REM dreams exhibit several general characteristics. They tend to include experiences of motion, some quite incredible; experiences that are visually vivid; experiences that are so bizarre that they are much like hallucinations; experiences in which persons, places, and times change helter-skelter, though the changes tend not to strike us (in the dream) as unusual; experiences quickly forgotten (except when we awake middream); and experiences that are often highly emotionally charged, most typically in a negative direction.

This is quite a list, but it is well supported by evidence of three types. The first is clinical. J. Allan Hobson (1994) and other sleep researchers have spent years waking up willing subjects in REM sleep and asking them to please record the content of their dreams. Subjects are asked to record the events and persons that appeared in their dreams as well as the emotional valence of their dream experiences. And as Hobson so delightfully describes, the above list of dream features is indeed robust. The second source of evidence comes from brain imaging. We know from recent experiments that the occipital lobe, for example, lights up during REM sleep with greater energy than it does when we are awake and actually looking at something! We also know that the amygdala, known to be integral to a great deal of emotional processing, is similarly hyperactive during REM sleep. The neuroscientist Jaak Panksepp (1998) aptly describes the observed hyperactivity in the occipital lobe and amygdala during REM sleep as comprising "neural storms." No wonder then the visual vividness and emotional charge of REM dreams. The third source of evidence comes from the study of neurochemistry. There are two neurotransmitters particularly crucial to our waking capacities for focusing attention, for reasoning in a stepwise manner, and for forming new memories—serotonin and neroadrenaline. During REM sleep, both transmitters, we now know, are turned off. The absence of these chemicals helps explain why our dreams shift from one unrelated scene to another and why the experiences of our dreams leave us with no new memories. Scenes shift madly because our attention mechanisms are disabled; we lose the ability to impose stability and order on our thoughts. And the scenes of our dreams leave no new traces in memory because the required shower of chemicals is likewise dried up.

It is significant that Hobson does not begin his inquiry with the assumption that the contents of our dreams are possessed of symbolic meaning. He does not try to analyze inward from the "meaning" of dream experiences to low-level mechanisms. Instead, he begins with the list of recurrent features (described above) common to REM sleep dreams and inquires after the systemic function of mechanisms that implement those features. He asks, for example, why the experience of motion is so prominent in REM dreams, and having analyzed into the large-scale activity of the brain (revealed by imaging studies) and some of the specific chemical mechanisms involved (revealed by the study of neurotransmitters), we now know that the neural circuitry that implements our capacity for *actual* motion is active during REM dreams. Why might this be the case? What systemic function is served? A plausible answer, framed in light of the directive in (A), is that the dreamed episode of motion is the reactivation of the neuronal patterns that underwrite actual motion, where the reactivation of these patterns serves the function of improving the organism's capacity for actual motion in the future. This function can be fulfilled without the formation of any new, consciously accessible memories of our dream experiences. It is enough that the neural circuitry implementing our capacity for actual motion is subjected to virtual rehearsals. The happy result is that when we awake, the nonconscious processes that constitute our capacity for motion are better tuned to the demands of the environment we inhabit.

A similar suggestion has been made regarding the emotional valence of REM dreams. It is widely hypothesized that we have a memory system dedicated to recollecting particular episodes or events in our lives—so-called episodic memory. Episodic memory is usually contrasted with semantic memory, namely, the recollection of general facts. My memory of the details of my daughter's birth is episodic; my memory that my daughter was born on Thanksgiving Day in the year 2000 is semantic. Now consider two facts. The first is that recent episodic memories—those accumulated over the last several days—appear to be stored mainly in the hippocampus. The second is that fragments of these recent episodic memories tend to occur in REM dreams. We ought to wonder why these two facts obtain, especially since, as I have said, the experiences we have during REM dreams do not form new, consciously accessible memories. Why, if no new conscious memories are formed, do we dream of remembered episodic bits and pieces of days recently gone by? What systemic function is served?

The answer given by Hobson, Panksepp, and others is intriguing. These theorists assume that a fundamental ingredient of our personality—an ingredient implemented in the lower part of the brainstem, it seems—is what we might conceptualize by analogy to the notes of a musical scale as an emotional register. Virtually all our experiences, whether awake or dreaming, are animated or colored by a range of affective responses to things in the world or to certain sorts of internal stresses. Hobson makes explicit the analogy to music:

> Genetics probably sets an individual range of emotions within us, just as it sets the range of our pulses and blood pressures. We are the particular people we are in large part because of the way our brain's emotional registers are set. We can tune this register to a degree as we mature from childhood, but the emotional music that is played out of us as personality is as distinctive as our fingerprints. When the volume is low, we are the only ones who hear this music. But turn the volume up and the whole world knows what state we are in. The amygdala is also like a player piano in that it runs through its own scores every day. Dreaming is an emotional recital: all stops are out and the keys are hammering away. (Hobson 1994, 156)

We can leave open the question of plasticity, of how far the register may be altered in the course of development or therapeutic intervention, for even if our emotional register is far more plastic than Hobson thinks, it is nonetheless plausible that a register with at least some standing emotional dispositions is indeed a fundamental part of our psychology.

With this assumption in place, we can explain the co-occurrence of intense emotional content and the fragments of recent episodic memories in REM dreams in terms of two straightforward systemic functions. The first is to integrate recent episodes, especially those with their own emotional salience, into the workings of our emotional register. The persons, places, or events in our recent experiences are assigned an emotional salience, and this has the likely effect of preparing us to react in the near future to those same persons, places, and events with increased efficacy. The increased efficacy occurs, of course, at the level of nonconscious processing. That is, although none of the experiences in REM dreams lead to the production of new memories that we can retrieve while awake, our nonconscious emotional orientation toward significant elements of our environment is fine-tuned in light of our standing emotional dispositions. This orients us toward our loved ones, our enemies, our colleagues, our fellow citizens, and so on, in ways we tend not to notice but in ways that color and affect our reactions to them. The second systemic function is plausible only if our emotional register can be altered in light of our experience. If such changes are possible, then the second function of REM dreams is to alter some element of our emotional register in response to particularly charged episodes. Experience may have the effect of changing the intensity of our emotional dispositions or perhaps expanding or contracting the range of dispositions we possess. Both functions, if they exist, are clearly anticipatory in nature.

The hypothesis that our dreams are anticipatory in this way is elaborated by Panksepp:

REM allows the basic emotional circuits of the brain to be accessed in a systematic way, which may permit emotion-related information collected during waking hours to be re-accessed and solidified as lasting memories in sleep. REM periods may allow some type of restructuring and stabilization of the information that has been harvested into temporary memory stores. During REM, neural computations may be done on this partially stored information, and consolidation may be strengthened on the basis of reliable predictive relationships that exist between the various events that have been experienced. The dream may reflect the computational solidification processes as different emotionally coded memory stores are reactivated, and the web of associated relationships is allowed to unreel once more and to coalesce into long-term memories and plans, depending on the predominant patterns of re-evaluation. The highest statistical relationships between events that regularly reappear together may be selected as putative causes for emotional concerns in the real world, and hence stored in long-term memory to be used as *anticipatory strategies on future occasions* when similar circumstances arise. (Panksepp 1998, 139; my italics)

That something like this occurs during REM dreams is, according to Panksepp, widely recognized among sleep researchers.

To strengthen the suggestion, Panksepp offers a tantalizing speculation and a few observations. The speculation is that REM arousal mechanisms—mechanisms that activate the brain during REM sleep and produce conscious experiences during REM dreams—may well be the mechanisms that, in our ancient ancestors, constituted the entirety of their waking lives! The mechanisms that underwrite our most florid and most bizarre dreams may well be descendents of mechanisms that provided waking consciousness for some distant ancestor. This speculation is plausible insofar as the mechanisms involved in REM arousal are located in parts of the brain that are evolutionarily primitive, the lower brain stem in particular. With this speculation as background, the hypothesis that REM dreams are anticipatory strategies appears further supported by the following observations. First, the amount of REM sleep an organism has increases after waking experiences involving new environments or stressful situations. Second, animals (including us) experience more REM when they are young than at any other time of life; the central elements of the emotional register are being formed or reinforced from the onset of development.[9] Finally, when the mechanisms that induce paralysis during REM sleep are surgically damaged, animals that are normally immobilized during REM begin to act out their dreams. The results are amazing, providing us "our clearest porthole into the nature of animal dreams" (Panksepp 1998, 140). This experiment has been performed on cats, and what we observe are dreaming cats rehearsing all of the behavioral strategies employed when they are awake! That REM dreams are anticipatory strategies is a compelling hypothesis with substantial theoretical and empirical considerations in its favor.

As in the case of vision and memory, discoveries concerning low-level mechanisms involved in dreaming force upon us a reconceptualization of relevant high-level systemic capacities. The revision here, however, is particularly radical. Having framed the inquiry in terms of (A) and having analyzed into the neural bases for dreams, it is no longer plausible to view our dream experiences as coded clues concerning the condition of our soul. Some of what occurs in REM dreams are fragments of recently remembered episodes, to be sure, but the meaning of those fragments cannot be read off the contents of the dream. If Hobson and Panksepp are right, the fragments of remembered episodes that occur in REM dreams are interacting with various elements of our psychology, including the standing features of our emotional register. The affective significance of the events and people recalled is integrated with the goals and motives of our personality. It may also be the case that elements of our personality are strengthened or diminished in the process. And all of this, of course, is a complicated process to which we have no conscious access; it is entirely nonconscious. Our dream experiences, therefore, are not introspective windows to our souls, our psyche, our repressed fears or lusts, because they are not windows to anything. They are the by-products of neural systems endowed with anticipatory functions that we are only beginning to detect. The reconceptualization is radical indeed.

If we step back from this handful of specific cases and consider more generally the joint power of (EH), (DP), and (A), and if we do this with the directive concerning conceptual change also in mind, we should, I think, heartily endorse the following directive for inquiry into our conscious capacities:

Nonconscious mechanisms (NC): For any conscious capacity of mind, expect that we will correctly understand that capacity only if we (1) frame our inquiry with plausible assumptions concerning our evolutionary history, (2) formulate competing hypotheses concerning the affective or cognitive capacities involved, and (3) analyze inward until we discover low-level, nonconscious, anticipatory mechanisms implementing the hypothesized capacities.

This directive assumes that our prior understanding of conscious mental capacities is probably incorrect. It also assumes that we are unlikely to correct these misunderstandings unless we analyze into the low-level, nonconscious mechanisms in the context of historically informed assumptions. The fundamental thought is that we cannot acquire knowledge of what we are like without achieving considerable distance from the infirmities of our first-person point of view. We must conceptualize ourselves as evolved animals and analyze inward in search of anticipatory, systemic functions, knowing full well that the functions we discover will likely force us to alter the way we understand the kind of animal we are. It is striking how deeply opposed this view is to the default assumption of all theology and most philosophy.[10] According to the progressive orientation—according to all my directives for inquiry, but especially the directive in (NC)—it is a grievous error to begin inquiry assuming that we already know what we are like with respect to our most important capacities as agents.

At any rate, on the strength of the examples discussed above, we should endorse the suggestion that the directive in (NC) also applies to the capacity that causes us to apply the concept 'purpose' to the parts of living things. There is no good reason to withhold the analogy. If mechanisms involved in vision, memory, and dreaming are possessed of anticipatory, systemic functions that force us to rethink the relevant conceptual categories, why not the mechanisms that incline us to "see" the parts of organisms as purposive? There are, so far as I can see, no good grounds for thinking that these mechanisms are unique in ways that would exempt them from what is known about our evolutionary history and the evident power of analyzing inward.

The question then is, What are the anticipatory, systemic functions of the mechanisms posited in the theory of mind theory? And the further question is, To what extent might the discovery of those systemic functions force us to reconceptualize our tendency to see and feel purposes where none exists? Lovely as they are, these questions remain largely unanswered at present. We have experimental evidence from Kelemen, Keil, and others that we are prone to conceptualize objects in our environment in terms of their purposes from a very early age, and we have the hypothesis that the cognitive and affective mechanisms that cause us to apply 'purpose' are the mechanisms identified in the theory of mind theory. What is presently lacking is a breadth of evidence at the cognitive and affective levels comparable to the evidence supporting the theory of mind theory, as well as evidence from neuroscience concerning low-level mechanisms.

A Fine Speculation

We can, however, cobble together speculations from disparate lines of inquiry. We may begin with the speculation that the capacity causing us to apply 'purpose' is a reliable detector of a specific type of self-motion. David Premack (1990) describes experimental evidence that human infants pay preferential attention to objects that appear to move by virtue of internal causes as opposed to objects that move as a result of being struck by something external (for example, another moving object). The suggestion is that one systemic function served by the capacity with which we apply 'purpose' is the detection of objects endowed with the capacity for self-motion. The relevant powers of detection are twofold. First, our attention is drawn to objects that appear to move by virtue of causes that are wholly internal and we are caused to conceptualize such objects as possessed of the capacity to move toward some internally established end.[11] Second, and taking a page from Kelemen's discussion, we are further caused to conceptualize the parts of the self-moving object as endowed with their own purposes relative to the ends of the self-moving object. We are caused to see self-moving objects as end-directed, and we are further caused to see the parts of those objects—those that contribute to the ends of the self-moving object—as functional or purposive.

Though these two steps apply to whole organisms, they may also be triggered by parts of organisms that exhibit their own signs of being self-movers. That is, some organismic parts may themselves exhibit motion that appears internally directed. Changes in the eyes—dilation and contraction, for example—may strike us this way. Or consider the expansion and contraction of the heart, the break down of food and elimination of waste, the reaction to sexual stimulation, and so on. Or consider the swimming capacities of human sperm. As science reporter Claudia Dreifus describes, it takes human sperm cells about fifteen minutes to swim to an egg after entering the female reproductive tract. Their built-in motors give them the appearance of being self-movers of a particularly determined sort. This appearance is most vivid when they reach and surround the egg. Here is a description reported by Dreifus from an interview with David Clapham, a scientist at the Harvard Medical School: "When sperm get to the egg, they need to crash through the ovum's membrane to deposit their DNA there. The way that happens is that at the end of its run, this ion channel [a channel that is utterly unique, found only in the testes and in the tail of mature sperm cells] brings the sperm calcium, which changes the shape of its tail and turns it into a kind of whip. The sperm is then propelled into hyper drive—pushing it into the egg with 20 times the force of normal swimming" (Dreifus 2007, 2). And note Clapham's appreciation for the apparent capacity of sperm to move themselves about in ways that appear highly purposive: "I'm fascinated by how determined [sperm cells] are ... each one seems an individual in the way they move. When they change from one motion to another, it's fascinating.... As you watch them under a microscope, you get the sense that they are going somewhere, or at least 'think' they are. They surround an egg and vigorously try to fuse with it. They don't give up until they run out of energy" (Dreifus 2007, 2).

The speculation is that we are, by virtue of our psychological constitution, highly prone to attend to objects that give the appearance of being self-movers and, moreover, highly prone to conceptualize the parts of those objects as being purposive relative to the ends of the larger system. This speculation apparently applies not only to organisms but also to the parts of organisms that are effective self-movers, even single sperm cells.

We are then, thanks to our evolutionary history, adept detectors of organisms and parts of organisms that exhibit signs of being self-movers. Or so it seems reasonable to suppose. We may further suppose that in our ancestral environment objects with the capacity for self-motion tended to actually *be* things capable of having purposes or ends with respect to us. And since things with purposes geared toward us may be friendly or hostile, it is reasonable to suppose that ancestors who were particularly good at detecting self-moving objects would have enjoyed some advantage. It is also reasonable to think that this capacity for detection provided an advantage only when paired with quick emotional responses that translated into immediate behavioral outputs, hence the value of an infant's preferential attention for self-moving objects and the early operation of an emotional register. More generally, objects in our ancestral environment with the capacity for self-motion tended to be living things capable of taking certain attitudes toward us, and that is why the entrenched capacity to mark other living things as potentially wishing to feed us, warm us, play with us, mate with us, or kill us is, in all probability, endowed with a systemic function that is aesthetically or affectively charged and very much oriented toward the future.

We might even speculate about the neural implementation of this systemic function. Antonio Damasio (1994) and his associates studied patients suffering from affective deficits caused by damage to the parietal lobe. In a revealing series of experiments, they compared the performance of normal subjects and patients with frontal-lobe damage on the following task. Subjects were seated across from an experimenter who controlled four decks of cards and who asked them to point to decks from which they wished to receive cards. Subjects were informed that each card indicated an amount of money. Some cards instructed the experimenter to pay that amount to the subject; others instructed subjects to pay that amount to the experimenter. Subjects were given a set amount of fake money (which nonetheless looked realistic) at the outset. The task of the game, they were told, was to try to increase their holdings or at least avoid bankruptcy.

It turns out, however, that decks A and B were relatively risky decks. When subjects received payouts based on these cards, the amount was relatively high (for example, $100). When, however, they had to pay in, the amount was disproportionately high (over $1,000). By contrast, cards in decks C and D paid less ($50) but also cost less ($100). The first crucial finding of these studies concerns the different strategies adopted by the two experimental groups. Normal patients, after discovering the very expensive payments required by A and B, shifted their attention to C and D. Despite the higher payouts in A and B, they decided that the extremely high payments required by A and B were too risky. By contrast, all the patients with frontal-lobe damage continued to request cards from A and B, acquiring greater and greater debt.

The second crucial finding became clear when skin conductance measurements were taken while subjects played the game, and what emerged is very much to the point of my speculation. It was discovered that as normal patients played, the responses generated by their autonomic nervous system, as measured by skin conductance, steadily increased when subjects chose a card from A and B. More important, as the game unfolded, skin conductance measures of normal patients showed an autonomic response in advance of—that is, in anticipation of—choosing a card from A and B. They showed an inductively acquired visceral aversion to the most risky decks. And, as you will have guessed by now, patients with frontal-lobe damage showed no such anticipatory response. They did not acquire an aversion toward A and B, and that, presumably, explains why they kept going back to A and B and drove themselves into bankruptcy.[12]

Might this be relevant to the capacity that causes us to attribute normative functions? Who knows? The analogy, however, is tempting. These experiments reveal that the human capacity for induction in contexts of uncertainty depends on implicit, nonconscious emotional responses. The undamaged parietal lobe is involved in the production of very quick emotional estimates of benefits and risks, estimates that exert profound effects on behavior. This is clear in the persistent patterns of self-destructive behavior that Damasio's frontal-lobe patients exhibit. The neural mechanisms involved accomplish all this without producing in us even a hint at the level of conscious awareness of what they are doing. This is not to say that normal subjects, if asked, would fail to notice that they feel a vague sort of apprehension toward decks A and B. But feeling unease is one thing; correctly interpreting one's vague feelings of apprehension is something else. And nothing available to conscious awareness would suggest to these subjects that, having executed a sophisticated bit of inductive reasoning, certain mechanisms of their brains were now serving a quite specific anticipatory systemic function.

The vague sense of apprehension to which normal subjects might have access is at least suggestive. The felt sense of purposiveness we feel when observing the systemic functions of the mammalian eye may be yet another vague feeling that we, at the level of conscious awareness, regularly misinterpret. The capacity that causes us to see or feel the eye as possessed of a functional norm may instead be possessed of an anticipatory systemic function that we, from the first-person perspective, cannot discern. We do indeed feel something when faced with functional marvels—our aesthetic sensibilities are undoubtedly engaged—but feeling, as I say, is one thing, and interpretation something else. The nonconscious systemic function of our conceptualizing capacity may be just as I have speculated, namely, to anticipate the behavior of objects that we perceive as self-moving, as directed by internal causes. And what we feel at the level of conscious awareness may be a partial or distorted glimpse of this systemic function. We may be caused to feel and think that the organism or organismic trait is directed toward some end, and that may be a misinterpretation of the visceral response that occurs when the relevant nonconscious mechanisms are causing us to anticipate the future movement of this organism or trait. It is at least plausible, in light of recent theories in vision, memory, and dreaming, to suspect that we are tricked in this way by our ignorance of the systemic functions of our own low-level mechanisms. The directive in (NC), at any rate, expresses an expectation that we ought to apply to our propensity to see the living in terms of purposes.

Conclusion

I conclude with two general implications. The first is that the joint effects of (EH), (DP), and (A), which I compress into (NC), cast doubt upon the alleged authority of our intuitions concerning natural purposes.

No longer can we take the apparent purposiveness of living things at face value. It is rational to divest the traditional concepts 'purpose' and 'function' of the authority they were once afforded.

Ruse's long discussion of the history of these concepts is offered as evidence of the importance of preserving these concepts. The psychological considerations described in this chapter, however, suggest that the history of these concepts is symptomatic of the power they wield in the economy of our psychology. The persistence of these concepts in our intellectual tradition, that is, reflects the persistence of psychological capacities that cause us to apply these concepts. In his second premise Ruse asserts what I have been suggesting throughout this chapter, namely, that we are compelled by the structure of our psychology to conceptualize living things in terms of 'purpose' and 'function.' But while Ruse thinks we must surrender to seeing living things as purposive even when we know they are not, I think we should at least attempt to free ourselves of this illusion. After all, Ruse and I agree that, while one part of our psychology inclines us to see the living as purposive, another part enables us to see that the living realm is devoid of purposes. So why give in to the capacity that we know leads us astray? Why not construct allies that might strengthen the other capacities involved, those with which we discern the truth? That, at any rate, is what my directives are meant to be—allies in inquiry.

The same point applies to the theory of proper functions. Like Ruse, advocates of proper functions take seriously the task of trying to preserve our concept of normative functions. They appeal to the admittedly powerful intuition that some traits retain their functional standing even when they are incapacitated and can no longer fulfill the functional task. The intuition we have that damaged or diseased items are malfunctional appears undeniable, and this seems to suggest that literal norms of performance belong to the parts of plants and animals. When we apply the directive in (NC), however, the gestalt of the situation is reversed. The best available evidence suggests that we are the kind of animal prone to see and feel purposiveness where none exists. The feeling we have that a diseased eye—one that can no longer dilate, say—nevertheless ought to be able to dilate is hardly grounds for constructing an elaborate theory of biological functions. It seems prudent to first check the source of our intuitions.

The second, concluding implication is that the directives illustrated throughout these chapters, when taken as an ensemble, exert considerable pressure against the instincts of conceptual conservatives. The problem facing the theory of proper functions, for example, is not that it flouts one or a couple directives. The problem is that it flouts all of them. And it is difficult to see how a theory that flouts them all qualifies as a successful naturalistic endeavor of any sort.

Notice first that the theory of proper functions fails to analyze into the mechanisms of natural selection. As we saw in chapter 5, the concept 'normative function' fails to analyze into the concepts and claims of the very theory that is supposed to vindicate these concepts. The lessons that motivate the directive in (EC) are simply ignored. Second, as described in chapter 4, the concept of normative functions is clearly dubious by descent; the directive in (DD) is ignored. Third, as we have seen in this chapter, the capacities that cause us to apply the concept 'natural purpose' are plausibly dubious by psychological role, for these capacities are entrenched and prone to false positives. The theory does not heed the directive in (DP). Fourth, as also suggested in this chapter, the theory of proper functions fails to take our evolutionary history to heart. It fails to conceptualize our psychological capacities as the anticipatory devices they are. And finally, the theory does not appreciate the directive in (NC). We should expect, on the basis of excellent historical grounds, that the high-level categories with which we conceptualize the capacities of living things are in error and that, as we analyze inward within an informed evolutionary context, our understanding of those capacities will change. This is particularly true of the capacity that causes us to conceptualize the parts of living things in terms of purposes. And the same five deficits afflict Ruse's view.

When we step back and view the effects of the ensemble as a whole, we are met with a striking scene. The theory of proper functions has the veneer of being robustly naturalistic because it appears, at least from a distance, to be a successful implementation of the concept location project. But upon scrutiny, the theory has nothing to recommend it to the progressive naturalist. Advocates of proper functions are conceptual conservatives who see themselves as naturalists only because they sometimes appeal to evolutionary theory. But being a naturalist is hardly the same as co-opting a handful of scientific concepts or claims in order to save one's cherished concepts. Being a naturalist requires a commitment to genuine exploration, to the sustained application of all the above directives for inquiry, combined with the expectation that concepts dubious by descent or by psychological role will be destroyed as new concepts are born.

If advocates of proper functions were progressive naturalists, they would be animated not by the hunches and instincts of a conservative but with felt doubts about concepts bequeathed to us by our theological ancestry and concepts that exert a particularly powerful sense of authority in the economy of our psychology. *That* would be their default position, and that would motivate them to ask how, exactly, the concept of normative functions analyzes into the mechanisms of natural selection. That fact of the matter, however, is that advocates of proper functions, along with Ruse, Chisholm, and Niebuhr, adhere to their own version of the default assumption. They assume we already know, in advance and at least in outline, what the world is like with respect to normative functions. On this view, the philosophical task is to specify what the world must be like in order for our concept of normative functions to be satisfied. That is to engage not in inquiry but in conceptual taxidermy. It is an orientation we ought to put behind us.

3

The Illusions of Agency

'Free Will' and 'Moral Responsibility'

SEVEN

The Death of an Aphorism: The Psychology of Free Will

All theory is against the freedom of will; all experience is for it. **SAMUEL JOHNSON, *BOSWELL'S LIFE OF JOHNSON (1791)***

After describing the retarding effects of our imagination early in the *Origin,* Darwin returns to this point throughout his discussion, only to drive it home one last time in the final chapter. The bulk of his concluding chapter is devoted to summarizing the many considerations both pro and con that bear on his theory. Still, in one of the last sections of the book, Darwin goes out of his way to repeat his earlier speculation on why his peers and people generally will resist his theory: "But the chief cause of our natural unwillingness to admit that one species has given birth to another and distinct species, is that we are always slow in admitting any great change of which we do not see the intermediate steps.... The mind cannot possibly grasp the full meaning of the term of a hundred million years; it cannot add up and perceive the full effects of many slight variations, accumulated during an almost infinite number of generations" (Darwin 1859, 481).

The chief cause of our resistance, he says, is a natural fact of the human mind, a natural infirmity of the imagination. This is to hypothesize that our failures of historical imagination are, in some instances, a consequence of failures in our psychological imagination, traceable to certain cognitive and affective limitations. The retarding effects of dubious concepts arise not only from our cultural history but also, and perhaps more fundamentally, from the constitution of our psychology.

When we turn from the study of life to the study of human agency, we are met with an even greater abundance of dubious concepts. The problems we face trying to understand our agential capacities are twofold. Much of our mental lives occurs beyond the reach of conscious awareness—that is the first problem—but, in addition, we are remarkably inclined not to believe that we are so thoroughly in the dark about our selves. We are inclined in fact to believe the very opposite, that we enjoy a high degree of first-person access to the contents and causes of our mental lives. That is the second problem. This double difficulty makes it prudent to consider whether the directives for inquiry employed in the study of life might bear comparable fruit in the study of human agency.

The aim of this final part of the book is twofold. I hope to convince you that we now know enough from psychology and neuroscience to appreciate the wisdom, perhaps the necessity, of employing my directives in the study of the human self. We have excellent grounds for thinking that progress in our knowledge of human agency is already proceeding in ways that parallel the growth of knowledge in twentieth-century biology. It is at least a good bet—our best bet, if I am right—that further progress in knowledge of our selves is going to require the determined application of these directives. Convincing you of this is the aim of the present chapter. But I also hope to convince you that we now know enough to draw the somber conclusion about human agency described in part 1. Although we are in a muddle about our own agency—although we do not know what kind of agents we are—it is clear that we are not the agents hypothesized by our Enlightenment and Romantic predecessors. And that means we are not the kind of agent that most contemporary philosophers persist in taking us to be. Convincing you of this is the aim of chapters 8 and 9.

The general argument of those final chapters has the following shape. There are, at present, two philosophical positions advocating a positive account of human freedom, namely, *libertarianism* and *compatibilism.* However, the best defense of libertarian freedom fails brilliantly the modest constraints of my directives (chapter 8), and recent defenses of compatibilism flout one or more of my directives (chapter 9). Since neither theory is plausible, we have in consequence no satisfactory account of human freedom. And if we have no satisfactory account of freedom—of what is thought to be a central capacity of human agency—then we are forced to conclude that at present we do not know what kind of agents we are.

The Collapse of Johnson's Aphorism

In the epigraph to this chapter, Samuel Johnson uses a mere baker's dozen words to describe the crux of the traditional free will problem. Free will is a problem for us because the experience of our own agency—the felt sense of being the author of our own actions—is as vivid as any experience that falls within conscious awareness. Indeed, some philosophers take our experience of agency to be a form of self-knowledge more secure than any scientific knowledge. Be that as it may, nearly all philosophers take our experience of agency to be a very central part of the phenomena that a theory of human agency must preserve. Yet this experience of authorship is hard to square with the concepts and claims of our best-confirmed scientific theories. Science is in the business of discovering mechanisms that make things work, while the experience of authoring one's own actions feels utterly unfettered and nonmechanistic. So the source of the problem is precisely this opposition between the way things feel from the agential point of view and the workings of the universe as conceptualized and studied by the sciences. This is the conflict Johnson so succinctly captures.

The intriguing question, however, is, What becomes of the opposition between theory and experience once we turn to the scientific study of experience? The conflict distilled in Johnson's aphorism pits science against felt experience, as if how things feel from the first-person point of view has the same epistemic authority as the concepts and claims of our best-developed sciences. That no doubt is how things stood in Johnson's day but—need it be said?—the problems that confront us today are different. The scientific study of the human mind, including the study of our experiences as agents, has changed dramatically the way we understand the reliability of our first-person experiences. The traditional antagonism between our humanistic and scientific worldviews is no longer a struggle between equally powerful opponents. My goal in this chapter, at any rate, is to convince you that the general shift toward a scientifically informed view of human experience is causing the collapse of the very framework described in Johnson's aphorism. We can no longer aspire to resolve Johnson's conflict, for it rests on an illusion the hidden causes of which scientists are beginning to expose.

Now, having read my warnings of the collapse of Johnson's aphorism, you may be wondering why I keep talking as though the experience of willing were relevant to the question whether or not we are in fact free. You may even regard the phrase "the experience of free will" as prejudicial or just plain oxymoronic, since the felt experience of authoring one's actions may obtain in a world with no freedom of will and free will may obtain in a world with no felt experiences. The worry more generally is that our felt experiences are accidental features of our psychology and perhaps our evolutionary history, while the reality (or unreality) of free will is a metaphysical matter independent of the vagaries of human psychology.

I agree that the nature of our experiences is by no means a reliable guide to the question whether free will exists. I especially agree that the vagaries of our psychology may induce us to feel and think ourselves free in ways we are not. I have, therefore, no intention of trying to settle the traditional question concerning free will on the basis of psychological theorizing. But none of this touches the real heart of the matter, which is this: traditional questions concerning human freedom have lost their former authority; the problem that dogged Johnson and others is no longer a genuine intellectual problem for us. Why? Because the only substantive reason we ever had for thinking that our agency conflicts with scientific theory or with metaphysical determinism is the persistent experience of consciously deliberating, choosing, and acting in the absence of any felt constraints. If we did not have this experience—if we did not feel our selves to be unfettered centers of command and control—the issue of free will would not be an intellectual problem for us. That is why it is crucial to begin with recent scientific revelations that appear to cast a dark shadow over the conflict described by Johnson.

I am assuming, of course, that whether a question or puzzle qualifies as a genuine intellectual problem is not a matter of choice. Nor is it a matter of chance or caprice. Genuine problems regarding the human self come from the inability, or the perceived inability, to make sense of ourselves relative to what we know about the world. They arise from the inability to see and feel ourselves in a way that is intelligible and comprehensive. And that is how things have stood for a long time regarding our experience of agency and the thesis of metaphysical determinism. But today, thanks to the scientific study of our felt experiences, the problems we face are different.

There is, moreover, the general point that even the most rarefied excursions into metaphysics are properly constrained by the facts of our psychology. More precisely, any metaphysical theory of the human self presupposes that we are in fact endowed with the sorts of low-level mechanisms required by the metaphysical theory. A theory of free will that fails to apply to us, that cannot be implemented by the mechanisms of our minds, may be a theory that applies to some kind of agent, but it cannot qualify as a theory that helps us understand the kind of agents we are.

Of course, you are (we may suppose) free (in some sense) to develop a theory that flouts this constraint. Perhaps there is some sense to be made of the attempt to articulate a theory of freedom that applies to any and all metaphysically possible agents. Perhaps. But that is to assume that our current conceptualizing capacities provide reliable access to the full range of metaphysical possibilities regarding 'agency'. It is to assume, more generally, a happy harmony between the capacities of our species and the fundamental nature of reality. And since we—our conceptual capacities and categories—are the products of a long, earthly, animal history on one rather puny planet, neither assumption is very compelling. Neither should be afforded authority in our inquiries prior to understanding the powers and the infirmities of our conceptualizing capacities.

What happens then when scientists interested in understanding human agency turn their attention to the felt experience of free will? What happens to the conflict upon which the problem rests? The short answer is that as psychologists analyze into the low-level mechanisms that implement our felt experiences of authorship and as they synthesize across related theories, it is increasingly clear that much of what we take ourselves to be experiencing is an illusion inflicted upon us by various infirmities of our minds. Our best available psychological theories diminish the plausibility of Johnson's observation. All theory is against freedom of will, and with recent progress in the sciences of the mind, less and less of our experience speaks in support of free will.

Undue Confidence: The Psychology of Naive Realism

The double difficulty described above comprises the following assertions: (1) Much of our mental lives is lived beyond the reach of conscious awareness. (2) We are inclined to feel and believe that we enjoy a high degree of first-person access to the contents of our mental states and the causes of our actions. The evidence for this double difficulty and the evident illusions it generates, as we will now see, demonstrate the value, perhaps the necessity, of my directives for inquiry.

To appreciate the first difficulty, consider a simple analogy. Unless you are a neuroscientist studying the human capacity for reading, you probably cannot explain the myriad processes taking place in your brain at this very moment, processes to which you have no conscious access. You consciously experience at least some of the wondrous effects of these processes—you grasp the meaning of this very sentence, for example, including the meaning of this sentence fragment that is nearing its end—but you know almost nothing of the causes producing these effects. The same holds for your capacity to process and interpret the meaning of utterances uttered by those with whom you converse. You grasp not only the propositional content of what is said but also the significance of vocal tone and inflexion, of facial movements and bodily posture. The same holds for other capacities we typically take for granted—proprioception, locomotion, vision, attention, memory, taste and smell, touch, and so on. The lack of conscious access to the causes of these capacities is not a matter of laziness or undeveloped skill but of architectural inaccessibility. So the force of the analogy is clear. If so many vital physiological and psychological capacities operate beyond the reach of conscious awareness, why expect anything different of the capacities that make us the kind of agents we are?

In the next two sections, I describe some of the nonconscious mechanisms that appear to implement our conscious experience of authoring our own actions. Before that, however, I want to consider briefly the second difficulty we face. Why, in particular, are we so inclined to think and feel that we know the reasons for the actions we perform? Perhaps part of the answer to this question is what Ross and Ward (1996) call our *naive realism.* Each of us, it seems, brings to the world and especially to social situations a set of assumptions, concepts, and convictions that frame the ways we react. Certain capacities of our minds generate very fast estimates of the significance of the situation in which we find ourselves, triggering a host of relevant cognitive and affective capacities. The hypothesis of naive realism is that this set of assumptions, concepts, and convictions determines the way we feel and think in the situation and, more to the point, inclines us to feel that our way of understanding the situation is neutral and objectively correct. According to Ross and Ward, the main elements of naive realism are three: (1) the assumption that we see things in an unmediated and objective manner; (2) the assumption that other rational and similarly informed persons will see things as we do; (3) the tendency to dismiss those who disagree with our view as ignorant, slothful, irrational, or biased. But what causes our minds to behave in these ways? The general suggestion is that each of us approaches a social situation with limited knowledge and extensive ignorance of what is going on. So each of us must solve what researchers in artificial intelligence call the "frame" problem by quickly generating an operable construal of the situation. We do this by imagining or filling in details that help us decide what is most significant about the situation we face. The origin of our naive realism then is that we generate a construal of the situation in the *absence* of a much-needed *check* on our construal capacities. There is, it seems, no check on the confidence we have that our construal is correct and, in consequence, no check on our felt confidence that our construal has been or will be adopted by everyone else. The genesis of our naiveté is an evident gap in the very structure of our psychology: nothing has the systemic function of reminding us that our particular construal is based on scant evidence gleaned from a particular perspective and that people with other perspectives and other evidence are likely to construe the situation differently.[1]

Why this absence? That is a lovely question about which we can only speculate. If our tendency toward naive realism is pronounced in social situations, we might do well to pair it with the theory of mind theory described in chapter 6. Our construal of the desires or purposes that motivate the behavior of other agents, thanks to our theory of mind mechanisms, may strike us with such force and immediacy that we are affectively inclined to trust them as accurate and disinclined to doubt them. Why the disinclination? Well, unbridled confidence in one's construal may have been of greater selective value than epistemic caution. Even today, feeling that one is right and freedom from felt doubt may conduce to decisive action, better learning, greater career prospects, enhanced interpersonal relations, and increased survival. Arrogance concerning one's self may be less costly than accuracy, perhaps by several orders of magnitude, especially in social interactions. Although this is an abstract speculation unconstrained by any specific historical evidence, it is at least suggestive regarding some of the most persistent and pervasive patterns of human behavior.

But whatever the correct explanation of our overconfident construals, the evidence that we are naive in this way is abundant and often amusing. In one frequently cited experiment (Ross, Greene, and House 1977), subjects were asked to walk around campus wearing a sandwich board commanding "Eat at Joe's." Subjects were also asked what percentage of other students would don the sandwich board. The results? Those who agreed to advertise for Joe's overestimated the percentage of others who would also agree; those who declined overestimated the percentage of decliners. That is, subjects in both groups overestimated the percentage of other students who would do as they did. This is the so-called *false consensus effect,* the tendency to overestimate the extent to which others will share our attitudes and preferences, our natural responses to objective reality. Further studies in Gilovich 1990 detected the same effect when subjects were asked about their preferences regarding pop music. Those who expressed a preference for music from the 1960s overestimated the percentage of others who would share their preference; ditto for those who preferred music from the 1980s. But Gilovich also discerned revealing hints of the causes of these different estimates. He found that those who preferred music from the 1960s generated in their minds a particularly agreeable sample of music from that decade and thought of less agreeable tunes from the 1980s. Those who preferred the 1980s did the same thing in reverse. The difference between the two groups was an effect, at least in part, of attending to highly selective exemplars. Hence the generation of biased and conflicting predictions.

The false consensus effect is the product of unchecked construals and serves as evidence for the first two elements of naive realism, namely, the tendency to think and feel that our view of reality is unmediated and objectively correct and, in consequence, that others will quite naturally agree with our view. The third element is, as Ross and Ward point out, a natural reaction to disagreement in light of the first two elements. If we feel confident that our take on a situation is the right one, we are apt to attribute defects of various sorts to anyone whose perspective does not jibe with our own. It is easy to see why we would be inclined to react this way. Of course it is tempting to dismiss those with whom we differ as victims of their own ignorance or sloth, but we also attribute defects to others even after a bit of well-placed reasoning. We are seduced in this way because we come to suspect, quite reasonably, that the judgment of others is biased. We all have, after all, more or less unique motives for what we believe and how we behave, and the observant observer will typically discern in others a positive correlation between some of their motives and the way they construe a situation. Such correlations are prima facie evidence of bias, of being unduly influenced by one's particular motives. We err, however, in failing to suspect ourselves of precisely the same sorts of biases. The problem, once again, is the evident absence of a much-needed check. Apparently nothing in the structure of our psychology—nothing in our affective repertoire—reminds us that our construals are likewise positively correlated with our motives. Nothing we feel serves as a check on our feeling that the situation we face is just the way it appears to us.

The absence of any such check permits us to persist in feeling that our construals are unmediated, objective depictions of the facts.[2]

Ross and Ward discuss several experiments that reveal the tendency to attribute bias to those whose construals conflict with our own. Perhaps the most troubling concerns disagreements among social scientists. As Lord, Ross, and Lepper (1979) report, proponents and opponents of capital punishment were asked to read a review of two studies about the efficacy of capital punishment as a deterrent to crime. (Subjects did not read the studies, only a review of the studies.) The studies discussed in the review employed different methodologies and offered conflicting conclusions concerning capital punishment. Although the review itself drew a balanced conclusion based on the relative merits of the two studies, the balanced nature of the review made no difference to the conclusions drawn by proponents and opponents—except to make them more entrenched in their own views. That is, each side fixed on the report that supported its own view and described it as superior to the other report. And both sides claimed to have found grave scientific difficulties in the opposing report! The attribution—even the confabulation—of biases could hardly be more vivid. Or more troubling.[3]

The truth of naive realism may help explain why we are so inclined to think and feel that we know the workings of our minds, including the reasons for which we act. The suggestion is that our constitutional inclination to overestimate the accuracy of our construals of social situations spills over to construals of ourselves. We have, after all, a more or less rich store of information about ourselves that we can access from the first-person perspective—by recollection, by imagination, by paying attention to how we feel or what we are thinking right now. And this apparent wealth of information may engender a felt sense of confidence that we know our own minds. Additionally, the absence of any kind of affective check on this confidence may serve to strengthen it. And if all this takes place in a social situation—if our minds are forced to generate quick construals of the situation that include an explanation of why we are acting as we are—then it is easy to see how the naive inclination to feel our construals as correct could lead to undue confidence regarding ourselves. It is easy to see how we might formulate and accept construals about our own reasons for acting even when they are woefully incorrect. The potential distortions generated by our naive realism are compounded when we consider the various low-level mechanisms that appear to implement our capacities as agents. As I will now describe, there are intriguing experimental and theoretical grounds for thinking that at least some of these mechanisms operate by producing in us the illusion of agential capacities that we do not possess. If so, then the retarding effects of naive realism are not to be ignored. Any tendency we have to multiply with undue confidence the power of our illusions is a tendency we must try to rein in.

The Mechanisms of Agency: Causal Interpretation

I begin with two simplified cases in which the reasons you give for acting a certain way are plausibly correct. Imagine that a stranger tries to snatch your purse from you. You yell "thief" while yanking your purse from her grasp. When you later explain why you acted as you did, you will probably get things right. Why? Because we typically bring to any situation a host of antecedent beliefs about the causal connections between motives and actions. Among this set is, presumably, the belief that a stranger who reaches for your purse is probably trying to steal it. Another is that having your purse stolen is an undesirable state of affairs that certain defensive maneuvers may prevent. So in the situation at hand, it appears likely that the reason you yelled "thief" and yanked your purse is that you were indeed possessed of the beliefs and desires just mentioned. This is plausible, at any rate, if no alternative explanation comes readily to mind.

Now imagine that your opponent in a public debate is someone you despise. Imagine further that, in the course of the debate, she makes a snide remark that misfires, providing you an enviable opening for response. When you later explain why you acted as you did—why you subjected your opponent to merciless humiliation—you will probably be right. Why? Because once again the situation appears relatively simple and unambiguous. You clearly desire to win the debate and rightly believe that one effective strategy for winning is making your opponent look intellectually flat-footed. There is also the generalization, true of many admirable persons, that humiliating your enemies brings immense satisfaction. In the situation at hand, it is thus quite likely that the reason you humiliated your opponent is that you desired and believed in these ways. This is compelling, at any rate, so long as no competing explanations present themselves.

In these cases, what makes it plausible that the reasons you give are probably the actual reasons for your actions? Just this: the *causal interpretive system* posited by Nisbett and Wilson (1977a) works best when the situation is relatively simple and unambiguous, rendering the possible causes of one's action few in number and transparent in character. Nisbett and Wilson posit such a system as part of the general hypothesis that the human mind includes a set of mechanisms dedicated to causally interpreting perceived events. And this is a compelling hypothesis. The human mind, like virtually all animal minds, includes some capacity or other to render the world causally intelligible. Whenever you perceive events or objects that are potentially related causally, your interpretive system is triggered to produce its best estimate of the causal relations involved. Navigating the world we inhabit would be impossible without some such ability.

But now consider what happens when you perceive yourself acting. When you act, you produce an event that includes your action. This triggers your interpretive system to generate an inference regarding the cause of your action, and the inferred cause, if it rises to conscious awareness, appears to you as the *reason why* you acted. By hypothesis, these inferential processes occur beneath the level of conscious awareness with breathtaking efficiency and speed that produces a noticeable absence of any effort on your part. The reason why you acted seems to appear (if it appears) quite suddenly in conscious awareness, and this absence of effort may encourage the thought that, when you give reasons for your actions, you are directly introspecting a reason that must have been present and efficacious prior to performing the action.

The interpretive system, more specifically, takes as input two types of information and treats them as premises in an abductive argument:

1. The perception of one's action in a given situation, and
2. The recollection from memory of causal generalizations (what Nisbett and Wilson refer to as our "a priori causal theories") relevant to the event and situation described in 1.

That is, your interpretive system receives as input the perception of you acting in response to certain situational features. Then, on the basis of these inputs, it performs calculations to answer the question, "What is the most likely cause of the action described in 1?" The question more fully is, "In light of the causal generalizations drawn from memory that appear most relevant to the situational features described in 1, what is the most likely cause of the action described in 1?" The system infers, in short, from a perceived *effect* to its most plausible *cause*—an inference to the best explanation—and produces as output the conclusion that

3. The cause of the action in 1 is such and such.

And as I have said, if the cause described in 3 rises to conscious awareness, then the agent takes this cause as the *reason why* he acted as he did.

That the human mind is likely endowed with a system that produces causal interpretations nonconsciously and with no apparent effort should not surprise us. When someone asks you for your social security number and you rattle off its nine digits with little effort, you are hardly surprised at your own prowess, even though you probably have no idea how your brain enables you to do it. If I ask you what George Washington's false teeth were made of, you may hesitate a moment or two as you perform a bit of conscious deliberation. In a matter of seconds you will probably remind yourself that modern dentistry is a recent development and that the materials used in the late eighteenth century were surely nonsynthetic; and on that basis you may venture a guess and say (correctly), "They were made of wood." Here too you will probably experience no surprise at your intelligent guess, even though the hidden neural processes involved are no doubt complex. If, then, it is credible that these and other capacities are implemented by highly efficient, complex, nonconscious processes, why think any differently about a system that implements one of our most elementary animal capacities, namely, the capacity to render the world causally intelligible and hence navigable?

And yet, though we should not be surprised, we should be taken aback by the implications for our view of human agency. What appears to us a simple act of directly introspecting reasons for our actions is, instead, an inference drawn by an interpretive system that not only works behind the scenes but also works *during* or *after* the fact. The system works, that is, during or after the action it explains. This is as it must be, since the information in premise 1 regarding the nature of the action does not obtain prior to the action, and the causal generalization required in 2 cannot be retrieved without a description of the perceived situation. And so, according to Nisbett and Wilson, the consciously accessible reason that we performed a given action—the reason we give ourselves or others—comes to conscious awareness during or after the action itself.

This is not to deny that we sometimes hold in conscious awareness, prior to performing the action, a thought about or intention to perform the action. Surely that is something we sometimes (though by no means always) do. And it may sometimes happen that our causal interpretive system, during or after the fact, infers a cause for our action that matches the content of our prior thought or intention. The claim, however, is that whether or not we hold in conscious awareness such prior mental contents, our interpretive system is in the business of producing its own causal inference during or after the action. Our system produces its own, distinct causal output irrespective of what was in conscious awareness prior to acting, and it is precisely this that opens up the possibility of substantial agential error. It opens the possibility that the agent, in attempting to understand his own actions from the first-person point of view, will come to believe that his action emanated from certain reasons when in fact it did not.

To see the potential for such error, suppose that a given action occurred with no prior conscious thought about the action. Suppose, that is, that our interpretive system produced a causal inference during or after the action; the actual, prior cause of the action operated beneath conscious awareness; and no conscious thought of the action occurred prior to its performance. It thus is possible that

A. Our causal interpretive system inferred a cause that matches the actual nonconscious cause of the action, or

B. Our causal interpretive system inferred a cause that does not match the actual nonconscious cause of the action.

But since the actual causes of our action are nonconscious in both cases, we are not likely to detect the difference between these two possibilities from the first-person point of view. This is one obvious source of error.

But now suppose that our action *was* preceded by a conscious thought regarding the action. It thus is possible that

C. Our causal interpretive system inferred a cause that does not match the prior conscious thought, or

D. Our causal interpretive system inferred a cause that matches the prior, conscious thought but does not match the actual nonconscious cause of the action, or

E. Our causal interpretive system inferred a cause that matches the prior conscious thought and also matches the actual nonconscious cause of the action (perhaps because the prior conscious thought caused or was caused by the nonconscious cause of the action).

It does not matter which of these possibilities occurs or occurs with greatest regularity. What matters is that if Nisbett and Wilson are right—if our minds are endowed with a causal interpretive system of the sort they describe—then when we engage in the activity of giving reasons for our actions, we are in no position to determine, at least from the first-person point of view, which of these possibilities is actual. We have no first-person, introspective grounds for claiming that possibility (E) obtains rather than (D) or (C), and this too is a substantial source of potential error. When we engage in the giving of reasons for our actions based on how things seem from the first-person perspective, we cannot justifiably claim to know the real reasons for our actions. We are faced with a potent epistemic defeater.

The only qualification to this defeater is, as I have said, the relatively simple and unambiguous cases like those described above. Why? Because in some cases it may be reasonable to believe that the inferred cause matches the actual nonconscious cause. In the case of the purse thief, for instance, it may be reasonable to believe that the possibility in (A) obtains. It is unlikely you reacted to the thief with the prior conscious thought to yell and yank as you did, but it is plausible, given the relative simplicity of the situation, that the actual nonconscious cause of your action matched pretty well the inferred cause. After all, the most salient features of the situation stand out—a stranger is abruptly reaching toward you and toward something you value—and that makes it reasonable to suppose that your interpretive system was triggered by the same features of the situation that triggered the nonconscious mechanisms involved. Aside from relatively simple cases such as these, however, the reasons we give for the actions we perform are indeed subject to a substantial epistemic defeater. Or so our best-developed theories suggest.

Why believe we are endowed with an interpretive system of this sort? There are three general categories of evidence which, taken together, constitute powerful grounds for accepting the hypothesized system. I will briefly discuss representative studies from each category, beginning with some described by Nisbett and Wilson (1977a). These studies illustrate a range of conditions under which subjects are in the dark about various facets of their own mental lives, including why they feel or believe as they do and why they choose or act as they do.

The first category comprises experiments of two sorts. Some experiments involve manipulations that alter the behavior of subjects in ways they fail to consciously notice, while others alter subjects' behaviors in ways they consciously recognize but the manipulations are accomplished by means of causes that go unnoticed, with the effect that subjects mischaracterize their own reasons for acting. I begin with two examples of the first sort.

In a study reported in Valins and Ray (1967), snake-phobic subjects watched a series of slides depicting snakes, interrupted by an occasional slide depicting the word *shock.* After viewing each "shock" slide, subjects received a small electrical shock. In addition, subjects heard a rhythmic sound while viewing all the slides. After the first couple viewings of "shock" and for the remainder of the experiment, the rate of the rhythmic sound was temporarily increased immediately after the "shock" slide. The rate of sound was not varied for any of the snake slides. Crucially, subjects in the experimental group were told that the rhythmic sound was the amplified beating of their own heart (they had been "wired" before viewing the slides), while subjects in the control group were told the truth, that the sounds were extraneous noise. What is striking is that after viewing all the slides, subjects in the experimental group—those who heard what they believed was a temporary increase in their heart rate after seeing the word *shock*—tolerated closer proximity to a boa constrictor than subjects in the control group. What explains this difference? Why did snake-phobic persons randomly assigned to one group suddenly react with less fear toward an imposing live snake than snake-phobic persons assigned to a second group?

The only difference, it seems, is that experimental subjects believed they were hearing an increase in the beating of their own heart in reaction to an anticipated electric shock. At the same time, these subjects also believed that their heart rates were unaffected by depictions of snakes. The explanation then is that these subjects were discovering—or so they thought—something about their own fears. They were discovering that they feared being shocked more than they feared snakes. And since they had evidently tolerated the shocks without incident, what was there to fear about the boa constrictor? The result was that these subjects apparently attributed to themselves less fear of snakes than they previously attributed to themselves. This change in self-attribution clearly affected their behavior.

And yet, when asked afterward about their fear of snakes, subjects in the experimental group showed no awareness of any change! They showed no awareness that their fear had been altered and no awareness that they had approached more closely an actual snake than they otherwise would have done. Of course, the lack of awareness in this case is not explicable in terms of our causal interpretive system, since the system was presumably not engaged—there was no perceived event requiring a causal explanation. Still, this study illustrates how our affective and cognitive capacities respond to cues from the environment and then alter our feelings and behaviors in ways we tend not to notice. It is no wonder we sometimes act without knowing the real reason why.

The same phenomenon is observed with respect to explicitly held views on social matters. Goethals and Reckman (1973) report a study of high school students who were questioned on their views about forced busing in the public schools as a means of achieving racial integration. Just two weeks after completing questionnaires, students were asked to participate in small-group discussions. In some groups, three probusing students were grouped together with a fourth student (an accomplice of the experimenters) who argued vigorously against busing; in other groups, three antibusing students were grouped with an accomplice who argued vigorously for busing. Afterward, students were given a different questionnaire to assess their current views on busing and, crucially, were also asked to recall their original views of the issue. They were reminded that the experimenters had their answers to the questionnaire from two weeks earlier.

The results are amazing. Most subjects reversed their consciously held views on busing—those who were antibusing before the small group now tended to support it and those who were probusing now tended to reject it—and, at the same time, most subjects insisted that their views had not changed as a result of the small-group discussion! Some subjects noted that their views had been broadened or strengthened by confronting issues on either side of the question, but not a single subject reported having changed his or her view. As in the snake phobia study, subjects of this study showed no awareness of a demonstrable change within their own minds. And since subjects were unaware of having changed their views about busing, they would have had no answer to the straightforward causal question—a question to which we know the answer—namely, Why did you change your position about busing?

A study by Nisbett and Wilson (1977b) takes us a bit closer to the distorting effects of the causal interpretive system. Half the experimental subjects watched an interview of a teacher who conversed warmly and with enthusiasm for teaching; the other half watched the same teacher talking coldly and expressing a lack of trust toward students. Afterward, subjects rated the teacher along two dimensions: (1) overall likability and (2) the attractiveness of his mannerisms, accent, and physical appearances. Of course, most subjects who found the teacher highly likable also found his mannerisms, accent, and appearances attractive, and those who did not like him found him unattractive with regard to these features. What is striking, however, is the apparent inability among subjects to correctly explain why they judged the teacher's features as they did.

Nisbett and Wilson asked two leading questions, each suggesting a distinct direction of causality. Some subjects were asked whether they thought their liking of the teacher had influenced their rating of his three attributes; others were asked whether they thought their liking of these three attributes had influenced their rating of the teacher. (As we will see, neither of these questions makes particularly good sense, and the fact that only these two questions were posed is probably important in understanding the results of the experiment.) What is striking is that all subjects denied that the teacher's overall likability affected their estimates of his mannerisms, accent, and appearances. Some even said that the direction of causality went the opposite way, that their estimates of these features determined how well they liked the teacher overall. Yet these latter responses are striking for their sheer implausibility. The three features itemized by Nisbett and Wilson were more or less invariant across both groups. Physical appearances, accent, and mannerisms were more or less the same whatever the teacher's demeanor; the most plausible direction of causality runs the very opposite to what subjects claimed.[4] The explicit reasons they gave for their judgments are demonstrably in error.[5]

What explains this striking lack of self-knowledge? What explains the fact that the reasons subjects gave for their judgments were not the actual causes of their judgment? The answer, I think, although a bit contorted, is at least consistent with the existence and effects of the causal interpretive system described above. Consider the situation of subjects who viewed the teacher expressing distrust toward students, and notice these two features. First, since all the subjects were students, it seems safe to assume that, after witnessing the teacher's avowed distrust toward them, they felt ill disposed toward him. Not toward his appearance or accent, but toward him. Why? Because when you are dependent on a person who occupies a position of relative authority, it is unsettling to learn that that person feels contempt toward you. Second, it also seems safe to assume that the students' felt aversion qualified as a psychological event in need of explanation. The question for their interpretive systems was "What about this man is causing my adverse feelings toward him?"

Surely the most obvious answer is "I feel negatively toward this authority figure because of his expressed contempt toward me and my classmates." Unfortunately, Nisbett and Wilson did not offer this as a possible answer. Worse still, the two questions actually proposed by Nisbett and Wilson probably encouraged subjects to feel that the choice was a simple either-or; subjects probably felt under some compulsion to choose one of the two options given. Just imagine your reaction if, during a psychology experiment, you were asked, "Do you find that man handsome because he looks good in green, or does he look good in green because you find him handsome?" Faced with these options, you are not terribly likely to go fishing for alternative explanations of your affective response to the handsome man.

Here then is what I regard as a plausible answer to the question, Why were subjects of this experiment so deceived about their reasons for acting? If the event to be explained is the felt aversion toward the teacher, and if the causal interpretive system is prompted (by the experimenters) with just two possible options, the causal interpretive system will select the one that makes the most sense or, as in this case, the one that makes the least nonsense. And in this case, since the event to be explained was the felt aversion toward the teacher—not toward his appearances or accent or mannerisms but toward *him*—the first possibility suggested by Nisbett and Wilson's leading questions is a nonstarter. It is a nonstarter because pointing out that we dislike someone is not an explanation for why we dislike that person. This, in turn, leaves their interpretive system with the second possibility, namely, that something about his specific features caused the negative feelings toward the man. Upon reflection, of course, this is not a plausible causal hypothesis (because of the point about invariance), but students participating in the study were not encouraged to engage in such reflection. They would have had, moreover, no particular reason to think that Nisbett and Wilson were plying them with implausible causal hypotheses; they would have found it natural to think that at least one of the options offered had to be correct. And that would have primed their causal interpretive systems to take seriously the options presented by the leading questions.

The second general source of evidence for the existence of a causal interpretive system comes from an experiment reported by Wilson, Laser, and Stone (1982) and discussed by Wilson (2002). The startling conclusion drawn from this experiment is that people who observe or read about our actions tend to attribute to us reasons for our actions that match the ones we attribute to ourselves. This appears to have been the case for subjects who, each day for five weeks, recorded two types of information: a description of their moods and a description of conditions such as weather, quality of relationships, quantity of sleep, and so on that plausibly affected their daily moods. At the end of the five weeks, experimenters calculated the correlations for each participant between the moods reported and the descriptions given of the weather, relationships, quantity of sleep, and so on. Furthermore, having turned in their daily logs, subjects were asked to estimate how much each of these conditions (weather, relationships, sleep, etc.) had contributed to their moods over the five-week period. In making these estimates, subjects were not allowed to look back at their daily ratings but had to make their best guesses without looking at what they had actually said. These latter estimates were then compared by the experimenters against the actual correlations between mood and conditions recorded throughout the five weeks.

The first result is that subjects were only modestly accurate in estimating the causal efficacy of these conditions on their moods. They wrongly estimated the efficacy of sleep, rightly estimated the efficacy of relationships, and produced unimpressive overall estimates. The second result is that estimates produced by a third party matched those produced by subjects who had studiously recorded their own moods for five weeks. That is, after subjects completed their five-week records of their moods, an entirely new group was recruited to estimate the causal efficacy of the same conditions (weather, relationships, sleep, etc.) on the daily mood of the "typical student" at their university. Nothing about the students in the first group was revealed to those in the new group.

Yet the estimates of those in the new group were just as accurate (and also just as inaccurate) as those in the first group. Wilson concludes, "The tremendous amount of information the participants [in the initial group] had about themselves—their idiosyncratic theories, their observations of co-variation between their moods and its antecedents, and their private knowledge—did not make them any more accurate than complete strangers" (Wilson 2002, 111).

Here then is evidence that information available to direct introspection does not play nearly as large a role in our reason-giving activities as we tend to assume. If third-party observers attribute the same basic sorts of reasons for acting as others attribute to themselves, we must at least try to explain the overlap, and one such explanation is similarity of psychological causes. The hypothesis is that the reasons given were the same because the same type of interpretive system produced the same abductive inference in response to the same perceived features of the situation. This is not to deny that information accessible only from the first-person point of view plays a role in the estimates we make of our reasons for acting. It is to say rather that whatever role introspection plays, it is not nearly as expansive as we traditionally assume. Introspective information, it seems, is swamped by information available to both first-person and third-party observers.[6] And that makes it plausible, as Nisbett and Wilson hypothesize, that our *reasons why* are the products of a system sensitive to culturally shared causal generalizations.[7]

The third general category of evidence comes from the well-known experiments reported by Gazzaniga and Ledoux (1978) on split-brain patients. One patient was shown a picture of a chicken claw in his right field of vision. That information was routed to the left hemisphere of his brain but was not communicated to the right hemisphere, since the usual connections between hemispheres had been surgically severed. He was also shown a picture of a snow scene in his left field of vision, which routed that information to his right hemisphere. In most people the right hemisphere has little or no capacity for generating language and the left hemisphere does the linguistic heavy lifting. After being selectively exposed to the snow scene and the chicken, the patient was presented with an array of objects accessible to both fields of vision (and hemispheres). He was then asked to pick objects from the array related to what he had just seen. Using his right hand (controlled by his left hemisphere), he pointed to a picture of a chicken; using his left hand (controlled by his right hemisphere), he pointed to a snow shovel. The experimenter then asked the telling question: he asked the patient to give *his reasons* for his choices. In response to the question, "Why did you pick those things?" the patient said, "Oh, that's simple. The chicken claw goes with the chicken, and you need a shovel to clean out the chicken shed."

Gazzaniga's explanation of this astonishing response is that the left hemisphere, the neural home of human language, has the systemic function of interpreting why we act as we do. What the split-brain experiments reveal is that when a person is asked for a reason, capacities implemented in the left hemisphere generate a causal interpretation based on information available to the left hemisphere. That makes it easy to understand why the patient picked the chicken: the left hemisphere had received information concerning a chicken claw and easily picked the chicken as most relevant. The revealing question is, Why did the left hemisphere say what it said about the snow shovel? After all, the left hemisphere did not see the snow scene and received no information about it from the right hemisphere. So why did it say what it said? The answer, according to Gazzaniga, is that the experimenters' question put the left hemisphere in an interpretive bind. The interpretive system was triggered when the experimenters asked the subject for his reasons; the system was called on to produce a causal inference, even though it was bereft of relevant information locked in the right hemisphere. The only information it had concerned a chicken claw and the left hand pointing to a snow shovel, and it used this information to get itself out of its interpretive bind by confabulating. It concocted a "reason" for the snow shovel by rendering coherent the partial information at its disposal.

There is further neural evidence for the misleading effects of a causal interpretation system from the Wada test. When the right hemisphere of the brain is selectively anesthetized, the emotions expressed by the organism are altered in a particularly striking direction. Instead of expressing emotions produced by our more primary emotional systems—fear, rage, lust, and so on—patients express relatively superficial social emotions. This suggests that the left hemisphere, home to Gazzaniga's interpreter, appears to be in the business of lubricating our social relations. Our deeper emotional capacities are not engaged. "Our left hemisphere—the one that typically speaks to others—may be more adept at lying and constructing a social masquerade rather than revealing deep, intimate emotional secrets" (Panksepp 1998, 302). If so, then our capacity for giving reasons appears to be responsive to an obviously restricted range of situational cues, which means that the potential for mistaken interpretations is large indeed.

I do not claim that these three sources of evidence establish the existence of the causal interpretive system posited by Nisbett and Wilson. I claim only that collectively they make it rational to postulate such a system; doing so helps explain a wide range of otherwise puzzling phenomena. Accepting the existence of such an interpretive system, however, does commit us to a critical claim: the capacities with which we apply the concept 'reasons for acting' are prone to a range of false positives. The postulated interpretive system tends to produce accurate causal inferences only when the situation is simple and unambiguous, and, by its very structure, the system produces inferences only during or after the fact. So the gap between the actual causes of our actions and the production of consciously accessible reasons for acting is too big and too entrenched for us to ignore. It virtually guarantees that, when the situation in which we are acting is difficult to parse, our "reasons for acting" will not match the actual causes. The directive in (DP) to withhold antecedent authority to concepts dubious by psychological role must be brought to bear. It must be applied to the way we understand what is typically assumed to be one of our most important capacities as agents, namely, our capacity to know and give reasons for our actions.

The Mechanisms of Agency: Apparent Mental Causation

The causal interpretation system just described generates causal hypotheses during or after actions we perceive ourselves performing, and that, it seems, is the underlying psychological mechanism with which human beings go about the business of giving explicit reasons for their actions.[8] There is, however, evidence of a distinct system that also plays a role in our reason-giving capacity. This system is not in the business of constructing causal hypotheses about why we act but rather creating in us the powerful feeling that we are the authors of those acts. The function of this system concerns not the *why* of our actions but the affective or visceral *who.* And, as we are about to see, the operations of this system can lead us wildly astray about the nature of our agency.

Daniel Wegner (2002) goes straight to the heart of my lead question, What kind of agents are we? His guiding question is, What causes in us the felt sense that we, with respect to some of our actions, are unfettered centers of command and control? What causes us to have experiences that, according to Samuel Johnson, convince us that our wills are free? Cast in more scientific terms, what are the environmental cues and the low-level psychological mechanisms that trigger this feeling in us?[9]

Wegner's answer is inspired by previous theories of the perception of causality, theories designed to explain why we perceive some pairings of events as causal and others as noncausal. We know, for example, from the experiments reported by Michotte (1954), that subjects report seeing a causal relation only if the timing between events is just right. If we are shown a ball rolling toward a second, stationary ball, and if contact between balls is immediately followed by the second beginning to move in the same direction, then we perceive the pairing as causal. If there is even a small time lag—if the second ball does not begin to move immediately upon contact—we do not see the pairing as causal. Instead, we see the first ball as having ceased its motion and we see the second as beginning to move on its own, presumably by virtue of causes from within. Cases in which the timing is just right illustrate part of what Michotte calls the *launching effect.*

Michotte's experiments are particularly relevant because he was able to generate the launching effect in the absence of actual contact. He was able to generate in his subjects the perception of motion transfer even when no transfer of motion occurred. He did this with an ingenious device constructed from two discs (discs A and B in figure 7.1), which he placed behind a screen that had a small rectangular hole cut from it. When disc B is set in front of disc A, two half-circles are depicted, one nested beneath the other, touching at just one point near the top of the half-circles (as in C). The half-circle underneath is crosshatched; the one on top is a solid black line. When both discs are rotated counterclockwise behind the screen, the top half-circle appears through the rectangular opening to be a solid object moving to the right. When the point of contact between the two half-circles appears, it looks as though a solid object is striking a crosshatched object, causing it to move to the right. The experiments performed with this ingenious device demonstrate the ease with which humans can be caused to perceive causal relations that do not exist.

Wegner's own theory, the theory of *apparent mental causation,* extends Michotte's theory. The suggestion is that we tend to see a pair of events as causally related to the extent we perceive the following conditions:

1. priority—the cause precedes the effect
2. consistency—cause and effect are semantically related
3. exclusivity—no competing, alternative cause

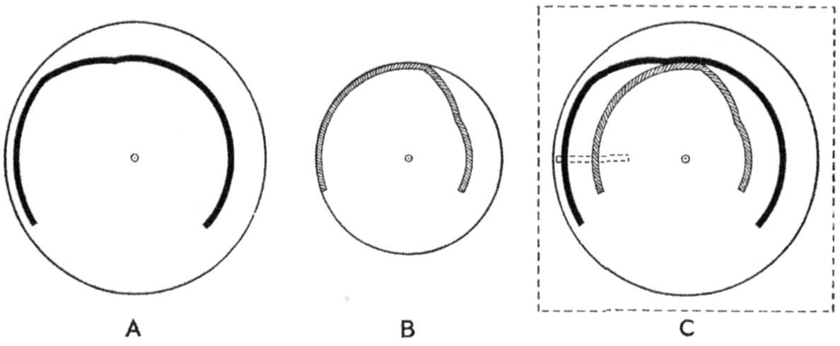

Figure 7.1 Michotte placed disc B in front of disc A so the two half-circles touched at one small area, as depicted in C. When both discs were rotated counterclockwise behind the screen in C at just the right rate, subjects reported seeing a solid object striking a crosshatched object, causing it to move to the right. This illustrates what Michotte called the launching effect (Michotte 1954).

The first member of the pair cannot occur too far in advance of the second; if it does, we do not perceive them as related causally. Nor can the first occur after the second. Moreover, both events must be conceptualized in terms that are semantically consistent. If I conceptualize one event as exercising my left arm by tossing a brick, and if I conceptualize a subsequent event as the shattering of the windshield of the car that cut me off a few miles ago (and to which I have just returned the favor), my perception of this particular pairing may feel less than robustly causal. Finally, the perception that the first member of the pair caused the second is diminished if we perceive a third event that might have caused the second. If, just as I launch my brick from the driver's window, I perceive my wife launching her own from the passenger window, I will likely be denied the felt satisfaction of having caused the resulting damage.

Wegner then applies these three conditions specifically to pairs of conscious events that produce the felt sense of authoring our actions, namely,

(the thought of doing action A) and (the perception of oneself doing A)

If I think of lifting my wineglass for another sip a few seconds before lifting and sipping, I come to believe and feel that my thought caused my action. That is, the capacities that enable me to perceive causality generally are triggered by my perception of this pair of events. Once triggered, these capacities cause me to think that my thought of having another drink caused me to have another drink. And, finally, this belief in the causal efficacy of my own thought causes a certain feeling in me. It causes me to feel that I consciously willed my own act, that the act of drinking was entirely my own. It is precisely this feeling—an emotion of authorship—that convinces us all that we are indeed, at least in some instances, unfettered centers of command and control.

Like Michotte, Wegner specifies conditions that reveal how readily we can be led astray, conditions that trigger the relevant psychological capacities and cause us to feel ourselves the authors of our actions even when we are not. More generally, the cognitive capacities that cause us to perceive causality are readily fooled. They make ready fools of us, as we happily think and feel ourselves to be rather impressive centers of command and control even when we are not. This is why Wegner calls it the theory of *apparent* mental causation—because the interpretive system that leads us to consciously experience our own thoughts as the causes of our actions appears to run on tracks distinct from the low-level, nonconscious processes that actually cause us to act.

The best evidence for the existence of these functionally distinct systems are cases in which our felt sense of willing fails to match the actual causes of our action. This happens in two ways. We sometimes feel that we have authored a given action even when we have not, and other times we do not feel we have authored an action even when we have. Wegner discusses multiple cases from both categories throughout his book. Begin with the feeling of having willed one's action even when something external to the agent is known (not by the agent but by others) to have caused it. Hints of this phenomenon are apparent in a range of priming experiments, where people's behavior is caused by stimuli of which they are unaware.

In experiments conducted by John Bargh and his colleagues, subjects are told that their ability to unscramble words is being tested. Those in the experimental group are exposed to a series of words typically used to describe elderly people, while those in the control group are not. Afterward, as subjects get up to leave the room, their posture and gait are observed; in other studies, memory tests are administered. The result is that subjects exposed to words such as *gray, retired,* and so on, compared with those in the control group, walked with poorer posture and slower step or scored lower on memory tests. This occurred even though subjects, when asked, detected no causal influence of the words they had read on their subsequent behavior. Presumably the mere priming of certain concepts—'gray', 'retired', and so on—triggers certain stereotypical concepts—'elderly'—which in turn produces downstream effects on cognitive and motor capacities. All this happens without our intending or even noticing it.[10]

Similar effects were discerned with respect to personal goals. Expose a person subliminally to the word *cooperation* and, as experiments have shown, that person is more likely than those in the control group to cooperate. Expose a person subliminally to words associated with rudeness, and a greater incidence of rude behavior is observed (Bargh, Chen, and Burrows 1996; see also Bargh et al.2001).The same occurs in response to nonconscious exposure to a host of additional concepts. And merely thinking about your significant other and the goals you share with that person tends to affect the goals you pursue even when you are not with that person. This happens to us without our noticing it at all (Fitzsimmons and Bargh 2003).

So nonconscious primes elicit effects on a range of behaviors—motor and cognitive—and also on our behavioral goals—the ends we find ourselves pursuing. These effects tend to go unnoticed by the very subjects affected. Were we to interrogate these subjects for the reasons on which they are acting, their answers would be in error. They would be the confabulations of Wilson's causal interpretation system or perhaps of Gazzaniga's left-hemispheric interpreter.

But now place the pervasive effects of such primes in situations where Wegner's three conditions can be manipulated. Begin with the first condition, the priority principle. Imagine yourself a subject of a psychology experiment, seated across from someone you believe to be a subject like yourself (but who is, in fact, an accomplice of the experimenter) and both of you attending to a single computer screen. You are asked to use a shared board perched atop a computer mouse to move the cursor to an object on the screen. The object is not specified, though there are several from which to choose. But the intervals at which you are supposed to move the cursor to an object are specified: you are to begin moving the cursor toward an object when you hear music over headphones and you are to stop the cursor when the music ceases. You are also told that, in order to assess the effects of auditory distractions, you will hear words spoken over headphones throughout the experiment. In fact, however, the words you hear name objects visible on the computer screen; you hear, for instance, the word *swan* and a swan is indeed sitting prettily in the depicted scene. Also unknown to you, during some trials the accomplice is instructed over her headphones to force the cursor to stop on a given object before the relevant interval has occurred for you. In these forced trials, the question is whether you will feel as though you have authored the action of stopping the cursor. More fully the question is whether, during the forced trials, your hearing the name of the object on which the cursor stops tends to increase your sense of having authored the stop, compared with forced trials in which you do not hear the name of the object just before the cursor is stopped. This is to focus on the efficacy of priority.

It turns out that hearing the word does indeed increase your sense of having caused the cursor to stop, even though the accomplice is the one who stopped it. We know this from the ingenious experiments reported by Wegner and Wheatley (1999) and fleshed out by Wegner (2002). The suggestion is that hearing the word *swan* primes the concept 'swan,' which in turn triggers your causal interpretive system to produce a causal inference. But, crucially, this priming effect is constrained by the priority condition. If you hear "swan" thirty seconds before the accomplice forces the cursor to stop, you feel no particular sense of will; if you hear the word a few seconds after the stop, the same result occurs. But if you hear it two or three seconds before the accomplice forces the stop, your sense of having authored the stop goes up; your interpretive system incorrectly identifies *your* prior thought about the swan as the cause of the action. By manipulating the priority condition in this way, Wegner and Wheatley trick the interpretive system to generate a false sense of will that feels altogether true.

Now consider cases in which an action is caused by a person who nonetheless feels no sense of will and even refuses to accept that the action is his own. This occurs in a surprising variety of ways, including apparent surges of creativity when scientists or artists experience a flash of insight. The insight comes unbidden as if from an external agent. This makes sense in light of the priority condition, since no prior thought about the solution occurred. There is also the remarkable range of automatisms described in chapter 4 of Wegner's discussion, including the bizarre cases of automatic writing and table turning. In nineteenth-century Europe and America, many people became convinced that spiritual intervention caused tabletops to turn or table legs to tap. A group of people would sit patiently with hands palms down on the table. Sometimes if they had enough patience the table would indeed begin to move. Participants typically felt no sense whatever of willing the table's antics; their felt experience seemed to indicate just the opposite, that the table was enacting the expressive gestures of some other agent, perhaps the spirit of a dead relative. Even after the distinguished scientist Michael Faraday demonstrated that the table was indeed turning because of pressure applied by participants' hands, enthusiasts of table turning remained entrenched. The absence of any sense of having willed the table's turning was taken as conclusive proof that some other agent was behind it all. To paraphrase Samuel Johnson, all experience was for the reality of spiritual intervention, even though science was against it.

The theory of apparent mental causation illuminates this case because none of Wegner's conditions is met. The exclusivity condition is flouted by the very nature of the activity; participants hope and even expect to be visited by some other agent. They expect someone other than themselves to manipulate the table in ways they can interpret. Neither priority nor consistency is met, since participants quite specifically do not intend to move the table. They may even intend to not move it. No doubt, some intentions come to mind as they sit before the table. They intend, presumably, to be patient, quiet, and receptive to the presence of mystery. And no doubt some participants hope to achieve some type of contact with a deceased loved one. But they do not intend to move the table by applying pressure with their hands. With respect to the act of moving the table, the question of priority does not arise and a failure of consistency is assured.

As striking as these cases are, the case of so-called *facilitated communication* is perhaps the most startling. It is certainly the most saddening. As Wegner describes in chapter 6, the Australian teacher Rosemary Crossley developed a clinical protocol to help people with impaired communicative capacities—people with autism or cerebral palsy—bypass their infirmities and at long last express their thoughts and feelings. Patients were seated before a keyboard or letter board, and facilitators, seated behind patients, were instructed to assist patients in the mechanics of responding to questions. They were told to keep the patient on track in various ways—by steadying the keyboard, supporting arms or wrists, guiding the motion of fingers, and so on. They were also told not to direct patients' responses or influence anything about the content of their responses. Their job was to help unlock the thoughts and feelings belonging only to the patients. And the results appeared miraculous. Patients who had never in their lives composed a single complete sentence were suddenly expressing intelligible and sometimes moving thoughts and feelings with well-formed sentences. Family members were understandably elated to learn that their loved ones were cognitively and emotionally intact and simply unable to communicate (up until now) for lack of effective expressive abilities.

The appearance of a miracle evaporated, however, as such appearances always do, when scientists performed a few simple experiments. Separate headphones were placed on patient and facilitator so that each person could be asked a different question. The result? Only the questions posed to facilitators were answered. In another experiment patients were shown certain objects while facilitators were not present. When facilitators returned to the room and patients were asked questions about the objects just viewed, none of the facilitated answers accurately described the objects. These and other experiments provide powerful evidence that facilitators, not patients, were answering the questions. These experiments also show how thoroughly confused we humans can be over the causal sources of our own actions. It is not plausible that everyone trained in facilitated communication was part of some grand conspiracy; the absence of any compelling motive should convince us of that. So something else, something about the workings of the human mind, explains how this could have happened.

The theory of apparent mental causation offers a ready explanation. As in the case of table turning, the exclusivity condition is not met. Facilitators come to their task believing or expecting or at least hoping that their efforts will result in the patient's expressing his or her own thoughts or feelings. There is the full-fledged intention that some *other* agent should do the talking, and this failure of exclusivity entails the failure of priority and consistency. The specific conscious intentions with which facilitators acted were presumably concerned with helping patients place their arms and fingers correctly. The most general sorts of intentions were probably along the lines of helping otherwise incommunicative patients communicate their thoughts and feelings. But at no point do the facilitators consciously intend to type out their own answer to the questions posed; if anything, they intend that that not occur. So it is easy to understand why, even after facilitators were given an experimental demonstration that they had indeed typed out their own answers to the questions, they nevertheless felt little or no sense of having willed the answers they in fact produced. The relevant sorts of prior intentions never entered conscious awareness, and the efficacy of some other agent was expected from the outset.

The theory of apparent mental causation, then, illuminates both types of cases—those in which a person feels a sense of having willed an action that was caused by something external to the person and those in which a person produces an action but feels no sense of having willed it. This is why it is plausible to think that the human mind is composed of two functionally distinct systems related to human action. One system takes as input our conscious thoughts about our actions and our conscious perceptions of having performed certain actions and produces as output the belief that our thought caused our action, along with the felt sense that we, or least some of our thoughts, are the centers from which we command and control our actions. The other system is what Wegner calls the *empirical will,* the set of low-level, nonconscious mechanisms that cause us to act. That these systems sometimes operate on their own, independent of one another, is a plausible conclusion in light of available evidence. And this means that the felt experience of having consciously willed our own actions is not a reliable indicator of whether we have in fact done so. The experience of authoring our own actions, the experience to which Johnson's aphorism appeals, can no longer be taken at face value.

Directives for the Double Difficulty

Wedding the work of Wilson and Wegner results in a particularly powerful defeater of our claims to know the reasons why we act as we do. The theory of apparent mental causation tells us that the way things feel when we are acting or immediately after we have acted is not a reliable indicator of the actual causal facts. It is not a reliable gauge of *what or who* caused the action. In particular, the feeling each of us sometimes has—the feeling that *I,* by virtue of a prior intention, caused my own action—is not to be trusted. It is not to be invested with the epistemic authority that Johnson gives it. And this, all by itself, puts each of us in an epistemic bind. If the theory of apparent mental causation is plausible, then we can no longer assume that the way things feel from the first-person point of view regarding the authorship of our own actions represents the actual causal facts.

There is a legitimate question about the scope of Wegner's central thesis. Is it that conscious thoughts or intentions *never* cause our actions, that *all* instances in which we think we consciously caused our actions are illusory? Or is the claim that only *some* cases are like this? Although this is an important question, it makes no difference for my purposes. Even if we grant that the causal efficacy of our conscious intentions is only sometimes illusory, the implications for our knowledge of our own agency are dramatic nonetheless. If we know that sometimes our feeling of authorship is an illusion, and if from the first-person perspective we cannot distinguish illusions of authorship from actual authorship, then we cannot justifiably claim to know in any given case whether we, by virtue of our conscious thoughts or intentions, actually authored our own action. Wegner's work, modestly construed, nonetheless constitutes a potent epistemic defeater of the experiences to which Johnson appeals.

But the psychological system that produces the felt experience of mental causation does not operate alone or without reinforcement, and the causal interpretation system described by Wilson is perhaps its closest ally. The system described by Wegner causes me to think and feel that I, by virtue of prior intentions, command and control what I do, while the system described by Wilson constantly reinforces these thoughts and feelings by generating for me the "reasons" for which I am acting. It does this in such a way that I feel as though I am simply discovering or calling up the same prior thoughts or intentions that I felt while performing the action, thoughts or intentions that appear to operate from the executive office of my command and control. And yet if the theories advocated by Wegner and Wilson are on track, the first-person, consciously felt effects of these systems are not reliable guides to the actual causal facts. We are faced with a doubly powerful epistemic defeater. We are potentially deceived after and potentially deceived during the action. Since we cannot distinguish from a first-person point of view whether a particular feeling of authorship or a particular reason for acting is accurate, it follows more generally that our felt experiences as agents are an unreliable guide to actual authorship.

***

Return then to the double difficulty described above. We feel confident we know the contents of our thoughts and feelings and the reasons why we act as we do, thanks to the fact that we are naive realists. At the same time, we are oblivious to the nonconscious mechanisms that generate in us the illusion that we are centers of command and control, thanks to a system that produces the feeling of conscious mental causation and the reinforcing effects of a system that produces causal interpretations of our actions after the fact. The force of the double difficulty we face in the study of our own agency could hardly be more pronounced, and that is why, in trying to understand our own agency, the importance of my directives can hardly be exaggerated. I take it that, by now, the expectation of conceptual change expressed in (EC) is clearly imperative to the study of human agency. I also take it that the directives in (DP), (EH), and (A) are similarly pressing. As the theories reviewed in this chapter make clear, some of the mechanisms of human agency produce in us unreliable beliefs and feelings about the actual causes of our actions. We must, therefore, in the interests of discovering the truth about our capacities as agents, deny authority to the very experiences that convince us (and convinced Johnson) that we are free and unfettered centers of command and control. We must withhold antecedent authority to the concepts in terms of which we conceptualize those experiences. There should be no presumption that an adequate theory of our capacities for deliberation, choice, or action must preserve those concepts. Hence the importance of (DP).

The significance of (EH) and (A) is especially interesting. We should hardly be surprised by the extent to which the workings of our minds are so sealed off from conscious awareness. Our ancestors could not have afforded a form of vision or memory encumbered in any way by the relative inefficiency of conscious processing; they would have been slaughtered in the struggle for existence. Why think any differently concerning our capacity for causal interpretation, especially in the context of social interactions? There is also the substantive assumption that our cognitive and affective capacities probably have systemic functions that are deeply anticipatory.

It is probable that for any capacity we wish to specify, the mechanisms that implement that capacity are endowed with systemic functions attuned to likely future events in the organism's environment. Instead of framing our questions about human agency in terms of conceptual categories inherited from our intellectual predecessors or categories that seem to describe our conscious experiences, we must inquire after the systemic functions of low-level mechanisms and processes. Doing so enables us to bypass the biases bequeathed to us by our tradition or imposed on us by the structure of our psychology. It also helps us reliably apply the directive in (DP).

What, then, are the anticipatory systemic functions of the mechanisms described in this chapter? What, if anything, is the most probable anticipatory, systemic function shared by naive realism, apparent mental causation, and a reason-giving system? One very general answer is that all three elements plausibly exert important influences on our capacities as social animals. This is, moreover, a suggestion offered by all the theorists discussed above. Wegner (2007), for example, discusses three specific hypotheses that address the evolutionary question with respect to apparent mental causation, namely, social signaling, social task allocation, and social control. In each case the systemic function hypothesized is anticipatory in nature. When we feel ourselves the author of our own actions, we can communicate to others and to ourselves what we take our selves to have done. The same system of apparent mental causation may permit us to treat our prior thoughts about actions as previews of what we are likely to do; this is information we can share with others or employ in predicting our own behavior. Moreover, the same system may be useful in providing us with some idea of the kinds of actions we can perform, which may contribute to the allocation of tasks within a society. Finally, the feeling of authoring one's own actions may serve as a guide to the sorts of actions over which one can exercise some degree of control and, in consequence, actions for which one can be morally praised or blamed.

The details of these proposals are less important than their general form. After all, the effects of naive realism clearly obtain in social settings; the effects of giving "reasons" are vital in social settings and would be rather puzzling in nonsocial ones; and feeling oneself the author of one's actions would be at least somewhat puzzling in the absence of social pressures,[11] especially if, as Wegner suggests, our felt sense of authorship is essential to the social practices of attributing moral praise and blame.[12] In general terms then, the anticipatory systemic function of these three systems is to lubricate the conscious capacities with which we engage in a range of social interactions. And this tells us something of great importance, namely, what the anticipatory, systemic functions of these systems most probably are *not.* The systemic function is not to endow us with free will, not to ensure that the "reasons" we give are the real reasons we act, not to ensure that we have an unbiased construal of the situations we face. No, the anticipatory, systemic function is to help ensure that our beliefs, feelings, and behaviors fall within the bounds of at least some forms of social order. Capacities that generate our illusions of agency endow us with the ability to anticipate how to behave in a range of social settings. They also help others anticipate how we are going to behave. And often enough the anticipatory effects of these capacities cause us to interact with others in ways that help maintain the social order.

This function attribution is admittedly speculative and partial. Even so, the anticipatory functions of a given capacity often diverge from our pretheoretical assumptions about those capacities. This is something we saw in chapter 6. Discovering the anticipatory systemic functions of our mental capacities reveals the extent to which we do not know what we are like. The same is likely true regarding the psychological systems described in this chapter, and the existence of speculative but plausible systemic functions for these systems only strengthens the suggestion that our pretheoretical assumptions about the nature of human agency are wide of the mark. This is something we discover by applying the directive in (A) in conjunction with the other directives discussed above.

Conclusion

Darwin thought that the failures of historical imagination were the effects, at least in part, of failures of psychological imagination, where the latter are caused by limitations or infirmities of our affective and psychological capacities. And, as the theories discussed above attest, Darwin was right about this. The failures of psychological imagination that afflict us when we try to understand our own agency are best encapsulated in the double difficulty described above. We feel sure we know the contents of our mental states and the reasons why we act as we do, yet we are blind to the nonconscious capacities of our minds that generate in us the illusions regarding our agency.

This, at any rate, is the conclusion we ought to accept for the time being. Of course further inquiry may force us to revise or discard this conclusion, but the balance of evidence available today favors it. And, as I have said, we need to decide at any given point in the history of human inquiry where to place our bets. The conclusion to draw regarding the nature of human agency is this: Johnson got it wrong, as did most of our theological and philosophical predecessors, concerning the problem of free will. A good deal of scientific theory is against free will and increasingly little of our experience is for it. And, as we are about to see, this has dramatic implications for contemporary philosophical theories of free will and for our view of ourselves as agents.

EIGHT

The Bare Possibility of Our Opinion: Libertarian Imperialism

To establish one hypothesis upon another is building entirely in the air; and the utmost we ever attain by these conjectures and fictions is to ascertain the bare possibility of our opinion, but never can we, upon such terms, establish its reality. **DAVID HUME, *DIALOGUES CONCERNING NATURAL RELIGION***

In light of what we are learning about the capacities that constitute the human mind, what kind of agents are we? Unfettered centers of control and command, as portrayed in much of our intellectual ancestry (and a good deal of our contemporary worldview)? Or are we something else? Moreover, in light of the directives for inquiry, how might we best approach the truth regarding these questions? How should we study ourselves?

In this chapter and the next, I address both questions by focusing on what is traditionally thought to be one of our defining features as agents. Contemporary theories of free will fall into one of two general categories, namely, those that deny and those that insist on the reality of human freedom. The latter category, comprising two general views that defend the reality of free will, shall be my focus. Although both views assert the reality of human freedom, they differ over its nature. Both parties agree that the felt experience of authoring our own actions does not readily cohere with the thesis of metaphysical determinism—Johnson's aphorism haunts us still. But, while *libertarians* insist that determinism must be false because the experience of unfettered freedom is so compelling, *compatibilists* insist that there are notions of freedom less demanding than libertarian freedom that are far more intellectually satisfying, in part because they are compatible with the apparent truth of determinism.

But, in my estimation, this entire way of framing the issue has lost its former intellectual footing. Not that the traditional problem has been solved, but that it no longer fits what we know about ourselves. We now know too much about the psychological and neurological mechanisms that bear on our capacities as agents to continue affording our felt experience of freedom the authority it once enjoyed. One half of Johnson's aphorism is dying and ought to be dropped from our intellectual worldview. Our most vexing problem at the moment is not whether determinism conflicts with our experience but, rather, the nature of our experience. We need to first discover the truth about our experiences.

The situations then is this: the stratospheric conflict between determinism and freedom has been superseded by the more earthly conflict between how we appear to ourselves as agents and the view of human agency emerging from our best sciences. The important intellectual problem of our day is the onslaught of conceptual confusion wrought by the fruits of scientific inquiry. If the felt experience of willing speaks less and less in favor of free will, then the problem we face is that we do not know what kind of agents we are. We do not know how to formulate a coherent, comprehensive view of our own agency. Though we are well equipped to see what kind of agents we are not, we are ill equipped to say what we are.

This has critical implications for any philosophical theory that asserts the reality of free will. In light of the conflict we now face, the pressing question is whether any philosophical theory that asserts the reality of free will is defensible. The question is whether the libertarian or the compatibilist offers a fruitful form of inquiry, whether there exists a positive account human freedom that retains its plausibility in the face of my directives for inquiry. The aim of this and the final chapter is to convince you that the answer is negative and, in consequence, that we are very much in the dark regarding the nature of our agency. I begin with the grand aspirations of libertarianism.

Romanticism and Ultimate Responsibility

Robert Kane's *The Significance of Free Will* is the most elaborate contemporary defense of libertarian free will, drawing on a broad range of resources that includes the long theological ancestry of the concept 'free will'; ordinary language claims from philosophers J.L. Austin and G.E. Anscombe; contemporary theories of nonlinear, dynamic processes; theories of quantum indeterminacies; the "hard problem" of consciousness; and conceptual categories inherited from our Romantic ancestry. I will have occasion to comment on most of these sources throughout this chapter. The most revealing feature of Kane's view, however, is the pervasive appeal to the Romantic ideal that we are creatures endowed with the capacity to create our selves, including the capacity to create features of our characters that make us "ultimately responsible" for some of our actions. On Kane's view, it is in the process of creating our selves that we make ourselves ultimately responsible for the characters we have and the choices we make. Freedom of will is both the process and the product of a lifelong artistic endeavor, the creation, as Kane puts it, of our own moral destinies.

With respect to the concept of freedom at the heart of his theory, Kane takes no prisoners. He wants everything—or nothing at all. His goal is to preserve the view of the human agent as a prime mover possessed of godlike capacities concerning the production of choice. "Free will," he declares, "in the traditional sense I want to retrieve ... is *the power of agents to be the ultimate creators (or originators) and sustainers of their own ends or purposes"*—where *ends* or *purposes* are choices that we normally express in the form of intentions (Kane 1996, 4). Stress falls on the status of the *will* and not the actions that emanate from the will. Kane concedes that there are perfectly good compatibilist notions of freedom that apply to the causal relations between an agent's will and actions, but he insists that there is an important concept of freedom that applies uniquely to the will, to the capacities we have for generating choices. And that is where he stakes his theory. Stress also falls on the capacity of agents to create, to be the prime movers of, elements that constitute the will: "to will freely, in this traditional sense, is to be the ultimate creator (prime mover, so to speak) of your own purposes" (Kane 1996, 4).

That we have these robust concepts of 'free will' and 'ultimate responsibility' and that we judge them essential to our status as persons is clear, according to Kane, from the relations these concepts bear to a range of other concepts. These other concepts include 'creativity' and 'autonomy' and refer to capacities that appear integral to living meaningful lives: "Among the creations we wish to be 'our own,' one stands out as particularly important—our own 'self.' Free will has been traditionally conceived as a kind of creativity *(poiesis,* in the language of ancient Greek thinkers) akin to artistic creativity, but in which the work of art created is one's own self. As ultimate creators of some of our own ends or purposes, we are the designers of our own lives, self-governing, self-legislating—masters, to some degree, of our own moral destinies" (Kane 1996, 81). The motivation for and significance of a libertarian theory of free will derive from the fact that we see ourselves as the creative prime movers of our own characters and, in some very substantial way, the artistic commanders and controllers of our own destinies.

What then are the conditions under which we are the ultimate originators of our wills? Kane gives an analogy to fiction. A novelist may begin writing with a more or less clear picture of the characters involved, but surprises inevitably occur. Conflicts arise that force her characters to respond in ways that are, well, novel, and often times even the writer is surprised. Often times her characters develop in ways that are unexpected and new. By analogy, human agents are sometimes faced with conflicts that evoke features of our characters that have hitherto been unknown. Other times, conflicts force us to cultivate our personalities in ways they have not been developed before. Sometimes we must dare to tread new ground. In doing so we discover what we are made of or what we are capable of.

In light of this analogy, Kane asserts that our capacity for self-creation is exercised in response to the felt resistance of conflicts we face. The obvious question then is, How does the process work? How does an agent equipped only with the constituents of her present self become ultimately responsible for the future self that her current actions help create? This question is crucial to understanding and assessing Kane's view. An agent, he says, is ultimately responsible for the operative elements of her will only if *she* is the creator of those elements, only if they are the products of a process orchestrated *by her.* It is *she* who is rightfully praised if things go well and blamed if they do not. So how does it happen?

To see Kane's answer to this question, consider his example of the woman who, on her way to an extremely important business meeting, witnesses a violent assault. She is gripped by the sudden emergence of conflicting motives. She wants to succeed in business and this morning's meeting is crucial, but she is deeply sympathetic and wants to render assistance. The conflict is pressing because, by hypothesis, she cannot do both; one of her motives is bound to lose out to the other. A life situation has triggered a motivational war, and now the woman, like the novelist, will simultaneously create and discover her own character. The sense in which she discovers something about her character appears plausible, since the conflict may be one she never faced before. Previously untapped capacities may come to the fore. What is less clear is the sense in which the woman, in living through this conflict, creates her own character in a way that makes her ultimately responsible for it. The only resources we have for influencing our future characters are the elements of our current character—or so it seems—and if we are not responsible for our current characters then we can never be responsible for our future ones. A nasty regress threatens. So the burden rests with Kane to show how the process of self-creation works and the sources of our ultimate responsibility operate in that process. So let us have a look.

Kane's Concept Location Project

The opening chapter of *The Significance of Free Will* gives a brief genealogy of 'free will' and 'responsibility' from antiquity to the present and, although Kane does not intend it this way, his historical sketch provides a vivid demonstration of the dubious historical roots of these concepts. Just as Ruse (2003) takes the lengthy genealogy of 'design' and 'purpose' as evidence that these concepts are significant and in need of preservation, the persistence of 'free will' in Kane's historical sketch appears to serve a similar motivating function. Apparently, we are supposed to be impressed by the staying power of this concept. At the very least, Kane nowhere suggests that the long theological history of the libertarian concept of freedom is a strike *against* trying to preserve it in present-day inquiry. As with Ruse, concepts dubious by descent are invested with undue authority; the directive in (DD) goes unheeded.

This is a telling objection against Kane's Romanticism. We know that the growth of human knowledge is retarded when we insist on preserving concepts dubious by descent. And the libertarian rendering of 'free will' and 'ultimate responsibility' are among the most dubious. They deserve as little antecedent authority in our inquiries concerning human agency as Niebuhr's 'self-transcendence' and Chisholm's 'agent causation.' Nevertheless, we cannot condemn Kane's view just yet, for two reasons. First, the application of (DD) is distinct from the simple elimination of the concepts involved, since even concepts dubious by descent may be vindicated by future inquiry. Except in cases where we already know that the concepts involved are hopeless—the concept of normative functions, for example—the progressive orientation requires that we hold fast concepts dubious by descent. Of course, the libertarian notion of responsibility is arguably the most hopeless of contemporary concepts, but Kane begs to differ. And this brings me to the second reason for delaying condemnation. Kane is offering what he regards as a decisive vindication of the libertarian concept 'ultimate responsibility.' He is defending the startling thesis that concepts with the longest, fattest theological roots have at long last been properly vindicated, and he thinks he can demonstrate this by engaging in an audacious version of the concept location project.

Kane claims to be a naturalist. He promises, at any rate, "to put the libertarian view into more meaningful dialogue with modern science" (Kane 1996, 115). He promises to do this in two ways. First, he assures us that, in the course of defending his libertarian theory, he will not adopt any metaphysical assumptions unique to libertarianism. Some assumptions are shared by compatibilists and libertarians alike, and only those, according to Kane, will he employ. He readily takes on assumptions about the nature of causation generally, for example, but pledges not to adopt any form of agent causation, including the view in Chisholm (1964). Second, he claims to locate central libertarian concepts amid the concepts and claims of contemporary cognitive science and physics. He agrees that an adequate theory of free will must be anchored in well-developed scientific theories concerning the actual world. Unlike Chisholm, who believed he could legislate what humans must be like given the alleged content of our concepts, Kane seems to think we must adjust our views of ourselves to fit what the world in fact is like. So we need to see where and why, in Kane's view, ultimate responsibility emerges in the process of self-creation. I begin with his notion of 'self-creation' and then consider how responsibility locates within that process.

Self-Creation

Kane begins by adopting the Romantic ideal of self-creation and then looking for scientific concepts and claims that appear to underwrite that ideal. The following is a list of ingredients required in the process of self-creation. In the case of the conflicted businesswoman, for instance, there exist

- two neural subsystems, one implementing each motive (one for her career, the other to render aid), both competing for dominance in the decision-making process.

In addition to such neural warfare, Kane posits

- a "self network" comprising the more or less standing motivational elements of one's character, against which the war is waged.
- genuine ontological indeterminacy within some neural molecules.
- the nonlinear, dynamic ("chaotic") amplification of the indeterminacy occurring in some neural molecules, producing systemic effects in the brain, including some that affect the neural processes implementing the agent's deliberation and choice.

One final ingredient, which operates at the level of conscious awareness but is implemented in the amplification of ontological indeterminacy, is

- the felt exertion of effort in overcoming resistance and resolving the neural war and hence the motivational conflict.

These ingredients, combined and baked just right, produce a lovely libertarian cake. Or so it is claimed.

This is how it works. Life events trigger motivational conflicts within the economy of our psychology, implemented by conflicts between distinct subsystems of the brain. Such conflicts are waged against the backdrop of more or less settled motivational dispositions that constitute one's character as it exists when conflicts arise. These dispositions within the self network influence the outcome of a conflict but do not by themselves determine the outcome. Other causal factors that come into play include the exertion of effort, presumably a conscious effort of the agent to insist on or to resist certain elements already in one's motivational repertoire, or the effort to decide on the basis of novel elements that arise due to indeterminism. There is, finally, the generation of novelty. This begins with undetermined subatomic events and their amplification. The indeterminacy involved is not epistemic or subjective, but entirely ontological. The claim is that certain subatomic events occur within molecules of the brain without being caused by anything. They just happen. And when they happen, a gap in the causal fabric of the universe opens up. Granted, gaps at the quantum level are hardly relevant to the macroprocesses in the brain that underwrite motivational wars unless quantum indeterminacies can be amplified. And, according to Kane, they can be amplified by nonlinear, dynamic processes. Kane does not tell us what distinguishes nonlinear, dynamic processes from other processes, but I take it that an uncaused event at the quantum level might produce several effects simultaneously that run in different directions within the neural molecule. Some of these effects, in turn, might produce several further effects that also run in many different directions at once, including some that extend outside the molecule to other mechanisms or processes in the brain.

Crucially, all of this activity—all the interactions among the listed ingredients—exerts a formative effect on the elements of one's self network. This is where the agent's inner struggle gives rise to self-creation. The agent, constituted as she is at present, helps resolve her own conflict by exerting effort. In doing so, she alters the relative weights of the elements in her self network or introduces elements that are altogether new. This happens in one of three ways:

1. the agent's effort reinforces elements in her self network that already exist;
2. the agent's effort weakens or destroys elements in her self network that already exist; or
3. the agent's effort introduces or cultivates new elements in her self network.

The most creative possibility is the third one. How might this happen? By chance, it seems. The uncaused event at the quantum level amplified by chaotic processing may introduce a bit of neural processing that is altogether novel. It may generate a neural possibility and, if it extends far enough, a psychological possibility that has never occurred before in the economy of this particular psychology. And if the agent, in exerting effort to end her conflict, settles the dispute by letting some altogether novel psychological factor win the day, then that new factor may take hold in her self network. If so, and if reinforced during future conflicts, the agent will have succeeded in making her own character in a way that is truly novel. She will have succeeded in being the prime mover of some aspect of her own will.

Nor should we overlook the importance of the first two possibilities, in 1 and 2. Although strengthening and weakening elements that already exist in one's self-network are not nearly as overtly Romantic as the introduction of something wholly new, adjusting the strength of elements in one's character is a creative act nonetheless. The alleged amplified indeterminacies can be contributing factors here as well. The businesswoman, though deeply sympathetic, may nonetheless have failed to act on her sympathies in the past; she may have been motivationally constrained by the very reasonable fear of retaliation for intervening. In the grip of the current motivational war, however, her standing desire to conquer this sort of fear may well come into play quite by chance. If, for example, the right sort of amplified indeterminacy occurs—if the woman is suddenly emboldened by a shot of adrenaline caused by the amplified effects of an uncaused quantum event—it may cause her to act in a way that reinforces her desire to conquer this fear. In this case, although no new element is added—the desire to conquer her fear has been a standing feature of her character—something novel, a small shot of courage, nonetheless occurs. In this way, she gives new shape to her self network by strengthening one element and diminishing another.

On Kane's view then, we have a taxonomy of agential devices operating at various levels, ranging from the quantum to the neural, from the neural to the cognitive, and all the way up to the phenomenological. The crucial steps in the process of self-creation are not simply those that lead to conflict resolution but those that enable the agent to reshape or expand her self network in the very process of resolving her conflict. Precisely this, the ability of the agent to alter the repertoire of motives that constitute the core of her character, is the originating source of our ultimate responsibility. We have the ability to create and re-create our own wills. We, like the great Romantic artists, must seek out and embrace the adversities of life, for these are the raw materials for self-creation. In doing so, we become both creator and creation of our own efforts. And we do so only when amplified neural indeterminacies give rise to possibilities not fixed by the past—there can be no freedom without genuine spontaneity.

The Locus of Responsibility in Self-Creation

These then are the basic ingredients and stages of self-creation. But the account is incomplete as it stands, for it is not yet clear how exercising the alleged ability to alter the elements of one's self network could ever redound to the credit of the agent. The obvious difficulty concerns the tension between the effects of amplified indeterminacies and the agent's efforts. If the agent's efforts do not determine the outcome of her conflict because of genuinely undetermined causal processes, then it is hard to see how *she* is any more responsible for the outcome than she is for the uncaused neural event. The amplified effects of an uncaused neural event are not the products of her present character but the products of a prior event with no cause. If, on the other hand, the indeterminacies do not undermine the contributions of the agent's efforts—if something in the agent's character bears the causal burden for the resolution—then the indeterminacies are idle in the sequence that leads to the action. And that makes them irrelevant to the question of responsibility. This is a dilemma that Kane and libertarians generally need to resolve.

Kane's response to the above dilemma takes us to the heart of his Romantic libertarianism. The agent is indeed responsible when, in the course of exerting effort to resolve her conflict, she chooses on the basis of the *right sorts of reasons.* Since Kane denies the truth of determinism, however, "reasons" cannot be understood on the usual backward-tracing model. The reasons that make an agent ultimately responsible cannot be exhausted by the elements in her character prior to the time of the conflict. Nor can the reasons be exhausted by the genetic and cultural endowments that produced her character. To insist on understanding reasons in this way is to beg the question against the libertarian view in favor of determinism. So Kane gives us an account of reasons that aims to render intelligible an agent's reasons for choosing without adopting the determinist's backward-tracing model. It also aims to make clear where and why the agent is the prime mover, the sole originator, of her own will.

Kane describes three distinct roles fulfilled by an agent's reasons, which he enumerates (r1)–(r3):

Whether the agents overcome temptation (and act on moral or prudential reasons) or succumb to temptation (and act on reasons of self-interest or for present satisfactions), the agents (r1) will in each case have *had* reasons for choosing as they did; (r2) they will have chosen *for* those reasons; and (r3) they will have made those reasons the ones they wanted to act on more than any others *by* choosing for them. (Kane 1996, 135)

The general claim is that all three reason relations obtain, despite the occurrence of amplified indeterminancies, and, moreover, when all three relations obtain, the agent's reasons for choosing justify the attribution of ultimate responsibility, despite the genuine indeterminacy in the process of self-creation. Let us consider each relation in turn.

The businesswoman has reasons for the choice she eventually makes because, by hypothesis, the deliberative conflict involves a struggle between two preexisting motives. No matter which course of action she chooses, she already has a reason in the form of a character-based motive for adopting that course of action. So (r1) is satisfied. The reason in (r2) is satisfied so long as the resolution to the conflict is causally influenced by the relevant motive. If the woman decides to rush past the scene of the crime to make her meeting, she will be choosing for the sake of her career so long as that specific motive is in fact causally influential in the resolution. Of course, the woman has other motives, but she does not choose for any of those unless they are efficacious in the actual causal sequence that leads to resolution. And although the reason for which she chooses does not determine the outcome of her conflict, given the indeterminacies involved, it raises the probability that the conflict will be resolved in accordance with that motive. And that, according to Kane, is enough to warrant the claim that she chose for that motive.

In explaining how (r3) is satisfied, Kane focuses not only on the woman but also on an alcoholic engineer tempted by bourbon in the face of job-related stress. In the following passage Kane quotes himself from a couple pages earlier:

Regarding condition r3, the relevant thesis is T29: "In cases of moral and prudential conflict ... both options are wanted" by the agents for different reasons, "and *the agents will settle* the issue of which is wanted more *by deciding.* If the businesswoman and engineer overcome temptation, they will do so *by virtue of their efforts.* If they fail, it would be because *they did not allow their efforts* to succeed. *They chose* instead to make their self-interested or present-oriented inclinations prevail" (Kane 1996, 135; my italics)

Despite the amplified effects of an indeterminate event that contribute causally to conflict resolution, the exertion of effort is the decisive contributing factor, at least when the agent's free will is exercised. More precisely, the decisive factor is whether the agent "allows" her efforts to succeed. The felt internal struggle is, according to Kane, implemented in the brain by the amplified effects of neural indeterminacies; the feeling of struggle and uncertainty is instantiated in objective indeterminacy. This objective indeterminacy does not void the agent's responsibility if the resolution achieved is a product of the agent "allowing" her efforts to succeed. To the contrary, it is precisely this capacity to "allow" one's efforts to succeed or fail that operates as the originating source of our acts of self-creation and, in consequence, as the source of our ultimate responsibility.

Here then is the connection to the libertarian account of reasons. The agent's capacity to "allow" her efforts to succeed or fail differs fundamentally from the determinist's model because not everything traces back to the current elements in one's self network. The agent's internal conflict is resolved in part by the causal effects of standing elements in her network but also in part by the amplified effects of neural indeterminacies that persist up to the moment of resolution. The crucial point is that the "allowance" capacities that enable the agent to overcome the indeterminacy and resolve her motivational conflict do not all trace back to standing elements of the self network. Instead, as the agent suffers the inner struggle and uncertainty, as the amplified effects rage at the neural level, certain elements in the agent's self network focus her attention on the task of resolving the conflict. This is the exercise of effort. And at some point the very same capacity, or perhaps a further capacity to focus attention at the metalevel, enables the agent to *allow* her efforts to resolve the conflict one way over the other. When this happens, the capacities of the agent responsible for the resolution cannot be traced back via a deterministic sequence to the agent's network; the effects of the indeterminacy that implement the agent's exertion of effort prevent any such tracing.

How do they prevent such tracing? Because the capacities of the agent to resolve the conflict are, by hypothesis, those in effect *at the moment of choice.*

Kane puts it this way: "Having plural voluntary control over a set of options implies being able to bring about *whichever* one (of the options) *you will, when you will* to do so" (134; Kane's italics). The crucial moment of choice takes place against the backdrop of one's self network but in the midst of genuine ontological indeterminacy. And it is at this moment that the agent, by allowing her effort to settle the conflict one way rather than another, acts as the originating source of responsibility.

We have here a "self-proclamation" theory of self-creation. The outputs we produce at these remarkable moments of choice are only partly traceable to our current character. What is also required, and what differentiates Kane's view from the determinist's, is that the most decisive causal factor in resolving inner conflicts is the capacity we exercise at moments of genuine indeterminacy to allow our efforts to go one way rather than another. These are the self-defining—the self-proclaiming—moments at the heart of Kane's Romanticism. It is as if the agent, by virtue of her effort, engages in the artistic act of proclaiming: "Henceforward I *shall be* an agent who chooses and acts on the basis of *this* kind of motive!" On this view, not everything that happens in the universe is settled by what has already happened. Some things—acts of artistic genius—occur because of our agential efforts in the here and now. The effortful struggle against adversity combined with the genuine ontological spontaneity of the present moment give rise to elements of our character for which we are ultimately responsible.

An Internal Dilemma

I am hardly alone in thinking that Kane's view fails in its aspiration to locate 'ultimate responsibility' in a nondeterministic universe. But I shall respond to Kane's theory in ways that other commentators have not, namely, in light of the directives for inquiry described in previous chapters. My aim is to show that according to our best-developed theories and best methods of inquiry, the libertarian view is without plausibility. This is true even if Kane's view were internally consistent and coherent, since the question whether Kane's theory adheres to the directives for inquiry asks for a great deal more than mere consistency and coherence. It asks whether we, subjects of the actual world, are agents of the sort he says we are.

Begin with the lessons that motivate the directive in (EC), especially the importance of analyzing inward. Just as the theory of proper functions fails to analyze inward from 'normative function' to the mechanisms of natural selection, so too Kane's theory fails to analyze inward from 'ultimate responsibility' to the concepts and claims of contemporary physics. To show this, I will not dispute Kane's assumptions regarding quantum and chaos theories. Even if all those assumptions are correct, his theory nevertheless fails to locate 'ultimate responsibility' within those (or any other) assumptions. It fails for the simple reason that the above dilemma concerning the locus of responsibility reemerges with respect to Kane's claims concerning the reasons for which an agent chooses.

Recall the businesswoman and the effort she makes. Kane says, "If the woman resolves her conflict by choosing to return to the alley and yell for help, *she* will have done so *by making the effort* to act morally, even if the outcome of her effort was not determined and hence not certain or guaranteed" (Kane 1996, 131; my italics). It is hard to see how this appeal to the agent's efforts avoids the above dilemma. If the amplified effects of neural indeterminacy contributed substantially to the agent's resolution, then it is arbitrary to attribute responsibility to the agent on the basis of her efforts. Her action, after all, is caused by a host of causal factors. If some of those factors are genuinely indeterminate, on what grounds do we privilege those that are not? If, on the other hand, the efforts or allowances of the agent do indeed justify the attribution of responsibility, they do so, it seems, only because they render moot the contributions of the amplified effects. But in that case the introduction of amplified indeterminacies is idle and the libertarian strategy falls down dead.

This, I gather, is where Kane would appeal to the ordinary language claims of J.L. Austin and others. Imagine Tiger Woods setting up for a three-foot golf putt. He has made the putt countless times in the past and failed rarely. This time, however, as he swings, a genuine ontological indeterminacy occurs. Some subatomic indeterminacy in Woods' neurology is amplified enough to cause a small twitch in his wrist. Still, the twitch is small, so Woods swings smoothly and sinks the putt. Following Austin, Kane challenges us to deny that sinking the putt is something that Woods did. It appears just obvious (says Kane) that sinking the putt was something Woods did and something for which Woods, the agent, deserves credit. Why? Because Woods was acting *for a reason* and acting *on the intention* of sinking the putt. That a genuine indeterminacy entered the causal sequence leading up to the act of swinging does not alter the fundamental fact that Woods was acting on and for his own reasons and, therefore, that the putt belongs to Woods. Or so our intuitions allegedly say. And something analogous is true in the case of the businesswoman. In exerting effort required to resolve her conflict, she acted for a reason and on the basis of a relevant intention. The occurrence of genuine indeterminacy does not diminish the fact that her reasons and intentions incline her toward a particular resolution without necessitating it. This, according to Kane, is how the exertion of effort implements responsibility despite the indeterminacies.

But this is not plausible. Note first that, even if Woods sinks a putt while his wrist twitches from an amplified indeterminacy, we cannot reasonably ascribe praise to Woods for the putt without investigating further the facts of his neurology and physiology. The first thing we need to discover is whether the amplified effects of the neural indeterminacy entered the actual causal sequence that produced Woods's swing. If there are multiple and parallel causal sequences that run from our brains to our motor capacities, then an indeterminacy in one need not affect others. So even if a twitch in the wrist can arise from a subatomic indeterminacy in the brain, the twitch may be irrelevant to the execution of the putt.[1]

But a more serious ambiguity looms. Suppose we discover that the twitch in Woods's wrist was indeed part of the sequence that produced his swing. Suppose that the neural indeterminacy produced dynamic, nonlinear effects that were among the actual causes leading up to the successful putt. We must nonetheless distinguish between the following cases:

C1: The effects of the amplified indeterminacy are such that had they not occurred then the same result—in this case, the same successful swing—would have occurred.

C2: The effects of the amplified indeterminacy are such that had they not occurred then a different result—in this case, a different swing (even if resulting in a successful putt)—would have occurred.

In the first case, effects of the indeterminacy occur in the actual causal sequence but make no causal difference in the sequence as a whole. The effects of indeterminacy function as mere noise within the process, canceled out by other causal factors, making no difference to the product of the process. In the second case, effects of the indeterminacy occur in the actual sequence and make a causal difference to the whole process by producing a different outcome. Crucially, if the counterfactual in C2 is true, then we have reasonable evidence that the effects of the indeterminacy *defeated* or *neutralized* the effects of Woods's talent and skill. We have reasonable evidence that, no matter what conscious reasons or intentions Woods may have had, we are *not* warranted in crediting him with the putt. Woods deserves credit for the putt no more than any of other factor in the causal sequence that resulted in the putt.

It must be the case then that C1 and C2 do not exhaust the options and that Kane requires some third possibility. Here is my best exegetical guess:

C3: The effects of the amplified indeterminacy are such that (1) without the agent's effort in resolving the conflict the result would be the product of genuine chance (the indeterminacy would make a real difference, à la C2) but (2) with the agent's effort in resolving the conflict the result would be settled (though not necessitated) by the agent.

But now the problem simply relocates to the second possibility in 2. If the agent's effort is what resolves the conflict, then the indeterminacy in C2 is rendered inefficacious by those efforts. Even if, at the very instant of resolution, there is a blitz of neural effects arising from an amplified indeterminacy, still, if the indeterminacy is resolved by what the agent allows, then the role of the indeterminacy in the resolution is effectively eliminated. The "allowance" capacities of the agent trump the effects of the indeterminacy. So Kane's libertarianism falls down dead.

Come at the point this way. If C3 is an accurate interpretation of what Kane says about the businesswoman in the passage above, then we must ask, *Who* is "allowing" or "not allowing" her efforts to succeed? This is not to beg the question against the libertarian by insisting that some elements of the agent's past or present character must determine the outcome. The question is about the very thing Kane insists on, namely, the exertion of effort at the very moment of conflict resolution. It is about the alleged capacity of the agent to bring it about that she is responsible for—that she is the originating source of—the choice she makes. The question is about the alleged capacity that casts responsibility for the resolution on the agent and not on something else. What is that capacity, what is its relationship to the indeterminacy, and how does it justify the attribution of responsibility to the agent?

It seems to me that Kane answers this question only by smuggling in a capacity of agents that operates at a metalevel. Over and above the conflict between first-order motives, the businesswoman apparently has the ability to conquer or surrender to one motive as opposed to another. She has a capacity to be, however briefly, a center of command and control. This is problematic in two ways.

First, it puts Kane's theory perilously close to alternative theories of free will that he explicitly eschews, including Chisholm's agent causation. On Chisholm's view, ultimate responsibility originates from an admittedly mysterious power of agents to initiate causal sequences that lead to actions without there being any antecedent efficient cause. Agents, according to Chisholm, can cause events but are not themselves caused by any events; they are little Thomistic gods capable of moving themselves without anything moving them.[2] Admittedly, the capacity that Kane attributes to human beings to "allow" their efforts to succeed does not appear bloated with the medieval metaphysical baggage that weighs upon Chisholm's view. But that is only because Kane is careful not to tell us what this "allowance" capacity comes to. Once we ask how any natural capacity of the human animal can ground the attribution of ultimate responsibility while preserving the efficacy of genuine indeterminacy, the air of mystery that attends Chisholm's agents quickly surrounds Kane's.

Second, no matter how Kane answers the question of who does the "allowing" or "not allowing," the alleged power of the agent to "allow" or "not allow" renders idle the amplified indeterminacies involved. By his own account of agential effort, the spontaneously produced novelties required for self-creation are swamped at the moment of creation by the remarkable capacity of the agent to allow or not allow her efforts to succeed. And if the agent's capacity to allow or not allow is the source of her ultimate responsibility, then, by Kane's own view, she must *already be* responsible for her allowance capacity. This shows that Kane's attempt to analyze inward from 'ultimate responsibility' to the ingredients listed above does not work. We have here yet another failed attempt at concept location, a failure that the directives in (EC) and (DD) would have prevented, had they been followed.

Kane's Conservativism

But suppose I am wrong. Suppose there is yet another vault concealed in Kane's labyrinthine theory into which we might escape. Even so, Kane must face the most serious defect in his view, namely, its conservatism, perhaps its imperialism, regarding concepts dubious by psychological role.

In trying to dodge the dilemma over responsibility, Kane insists that genuine indeterminacies at the neural level do not undermine the attribution of responsibility for efforts made at the level of *conscious awareness.* The agent is responsible for what follows from the conscious exertion of effort even when the causal sequence involved contains substantial indeterminacy. The agent makes her self responsible by issuing a proclamation that applies to the future self she creates. There is a kind of authority imposed on one's future self by the act of self-proclamation that takes place in the context of the spontaneity that comes from amplified indeterminacies.

One way that Kane argues for this claim is by analogy. He first asks us to imagine that the businesswoman, after resolving her conflict, is whisked off to a neuroscientist to investigate the brain processes involved in producing the resolution to the conflict. Lo and behold, it is discovered that the neural processes included amplified indeterminacies. So Kane asks,

> How should we respond to this new neurological information? Critics of indeterminist theories ... might insist that, as a result of this evidence, we should say that the resolution of uncertainty and conflict in the woman's mind was not really her "choice" at all, but rather something that merely happened to her ... She thought she was *choosing* or resolving the conflict, but now that the real happenings in her brain are known, we (and she) must say that she was mistaken. In response, I argued [in an earlier chapter] that there is no good reason whatever to accept such a conclusion from neurological evidence alone. We have no more reason to do so, I suggested, than to conclude that our thought processes and their conclusions are not really our doings if it should turn out that they involved some indeterministic chaotic processes in the brain, as well they might, according to some recent suggestions. (Kane 1996, 182)

The argument assumes that our thought processes—deriving a conclusion from a set of premises, for instance—are really "our doings" even if implemented by neural processes that include amplified effects of prior indeterminacies that reach all the way to the drawing of inferences. This premise is not defended by any argument; it is merely assumed. The argument further assumes that, by parity of reasoning, we should agree that a parallel point holds in the case of conscious efforts that result in choice. We should agree that the choices we make by virtue of conscious exertions of effort are, like our conscious thought processes, our "doings" even if the deliberative processes are the effects of amplified indeterminacies.

But this is not a good argument. One problem is that the analogy cuts both ways. If we have grounds for thinking we are not responsible for our conscious efforts when implemented by amplified indeterminacies, we are hardly likely to think that thought processes implemented by amplified indeterminacies are any different. At any rate, the above dilemma rests upon the general worry that *any* mental act implemented by substantive indeterminacies cannot redound to the agent's credit. So Kane's analogy runs headlong into the very objection it is supposed to resolve.

A further problem is that even if we pretend that the analogy has force, it nevertheless fails to resolve the dilemma. The challenge is to explain how an agent can be responsible for a decision when causal factors beyond her control—beyond the control of anything in the universe—played a substantive role in the production of that decision. Kane's response is to insist that it must be possible for responsibility to accrue under conditions of indeterminacy, for otherwise we would be forced to say that some of our thought processes are not really our own. But this says nothing about *how* it is possible that responsibility accrues under conditions of indeterminacy. The analogy to processes of thought reveals nothing about the sorts of processes that could underwrite responsibility in the face of indeterminacy. The burden surely rests with the libertarian to render intelligible the emergence of ultimate responsibility from indeterministic processes, to show how the mechanisms or processes that constitute the amplification of subatomic indeterminacies plausibly could implement moral responsibility.

But the most telling problem with the above argument is the extent to which it flouts the other directives for inquiry, especially the directives in (DP) and (NC). It affords undue authority to the apparent contents of our *conscious mental states,* especially those involved in exertions of effort to choose among conflicting motives. Recall the double difficulty described in chapter 7. That a great deal of our mental lives occurs beneath the level of conscious awareness is as clear as any thesis in contemporary psychology; that we are fooled into feeling that we have accurate conscious access to the actual causes of our actions is likewise clear. Anyone wishing to discover the truth about our capacities as agents must develop strategies that neutralize the retarding effects of this double difficulty. But Kane takes no such precautions. Worse yet, he proceeds as if the double difficulty does not exist. He proceeds as if it is just obvious that the first-person phenomenology we experience, including the "soul searching" turmoil triggered by motivational conflicts, is so reliable that it can serve as the originating source of ultimate responsibility. This is where Kane's conservatism spills over to imperialism. He embraces the default assumption of all theology that we come to the study of human agency already knowing what we are like, at least with respect to core concepts of agency.

That Kane is committed to the epistemic authority of our first-person phenomenology is clear in several ways. The first is that he openly endorses the self-proclamation view of self-creation. But another is the following passage in which he compares our sympathetic businesswoman to hapless Jane who, after consciously deliberating and deciding to vacation in Hawaii, inexplicably finds herself vacationing in Colorado. We are to suppose that, without any cause, the atoms in Jane's brain suddenly swerved, overriding her prior decision without any additional deliberation. In this case, Kane allows that the indeterminism involved in Jane's case undermines the attribution of responsibility. But

there is no legitimate reason to generalize from cases like Jane's and say it must always be so. Consider the businesswoman by contrast. Her experience, unlike Jane's, is of *consciously and voluntarily choosing* to follow her moral conscience and to return to help the victim, thereby resolving a preceding uncertainty in her mind. Also, in the businesswoman's case, unlike Jane's, the indeterminate process discovered by the neuroscientists immediately preceding the choice was *experienced by her* as *her own effort of will,* not merely as a random occurrence in her brain that happened to influence the outcome. Given these differences, it would be hasty, to say the least, to lump the two cases together and draw conclusions about the businesswoman's case from Jane's. (Kane 1996, 182–83)

Of course we should not draw conclusions about the businesswoman from cases in which a quantum jump not associated with any process of deliberation suddenly alters a prior decision. But who is suggesting we lump together cases like these? The challenge raised by the dilemma is not about cases in which atoms inexplicably swerve in the void and alter a decision already reached; it is about indeterminacies that are, by Kane's own avowal, partly constitutive of an agent's deliberations.

Although Kane is here attacking a straw objection, his discussion is revealing. Consider in particular the contrast he draws between the businesswoman and Jane with respect to first-person phenomenology. Kane insists that the businesswoman experiences "her own effort of will" as she "consciously and voluntarily" chooses to follow her conscience. No matter what the lower level neural processes, it is, on Kane's view, the experience of consciously exerting an effort of will that establishes the conclusion concerning responsibility. The consciously aware agent is the sole originating source of responsibility; responsibility accrues only when the effort of will is a product of the agent's own conscious and voluntary exertions.

Indeed, if we return to the early pages of Kane's discussion, we find him building this very point into the theoretical definitions he constructs. "Reasons" and "motives," he says, are psychological attitudes of the following sort: To be "correctly" or truly cited as reasons, psychological attitudes must play a role in the etiology of choice or action—they must influence choice or action in some manner or other that is not easy to specify. It is one thing to *have* a reason and another to choose or act *for* that reason. Jane may like river rafting and this may have been a reason for favoring Colorado, had she thought of it. But in fact, her mind was focused on skiing and visiting friends, and river rafting never entered into her decision. It was a reason she *had,* but not one *for* which she chose. (Kane 1996, 28; Kane's italics)

This, on its face, is a reasonable distinction. Jane likes river rafting, and had she thought about her desire for that activity then it would have been a reason for choosing to vacation in Colorado rather than Hawaii. But her desire did not enter into her conscious deliberations and hence did not become a reason for her to choose as she did. For Kane, then, something is a reason or motive *for* a given choice only if it was part of the agent's *conscious* exertion of effort. It needs to be among the reasons the agent would explicitly give for the choice she made.

This way of defining "reason" and "motive" is then incorporated into Kane's definition of voluntary action: "An agent acts *voluntarily* (or *willingly)* at [some time] t just in case, at t, the agent does what he or she wills to do ... *for the reasons* he or she wills to do it, and the agent's doing it and willing to do it are not the result of coercion or compulsion" (Kane 1996, 30; the last italics are mine). The focus here is on the nature of voluntary action. An action is voluntary only when the agent does what she actually chooses to do and only if the agent's reasons for choosing (inputs to the psychological system) are the salient causes of the choice made (the output). Kane thus defines the key term "voluntary action" in terms of the reasons for which a choice is made, that is, in terms of the motives that enter conscious awareness. An action is voluntary only if it results from the consciously accessible reasons for which the intervening choice was made.

But—need it be said?—Kane is wrong that our conscious reasons or motives for choosing or acting are, in general, accurate reflections of the causes of our actions. We often do not know why we act as we do and we often are fooled into thinking we know the reasons for our actions. Of course, there may be instances in which the reasons we give for our actions are indeed accurate—when, for example, the situation is simple and the causal generalizations we select correctly apply to perceived features of the situation—but such accuracy is out of the question in cases relevant to Kane's theory. By his own avowals, the process of self-creation, the process in which responsibility emerges out of amplified indeterminacies, depends crucially on episodes of severe motivational conflict. Kane describes these episodes as "soul-searching moments of moral and prudential struggle," periods of great stress in which "agents are torn between conflicting visions of what they should become" (Kane 1996, 130). As a rule, episodes in which we suffer such internal turmoil are not going to be simple in the ways described by Nisbett and Wilson (1977a). They are not going to be ones in which the potential causal factors that we perceive in the situation reflect the actual causes of our actions. Just the opposite is probably the case.

Wegner's theory raises an especially pointed problem. In order to avoid the determinist's model of deliberation and choice, Kane insists that the source of responsibility is the agent's exertion of effort at the moment of conflict resolution. When the businesswoman, by virtue of concentrated effort, allows this effort to select one motive over the other, responsibility accrues to her precisely because it was *her* allowance of effort that, at that very moment, resolved the conflict. Now consider what the first-person phenomenology of such moments of truth must be like. As Kane says repeatedly, they are experienced by the agent as *exertions of effort.* But as we saw in chapter 7, it is precisely those experiences, those moments of *felt effort,* that are called into question by the wide range of experiments supporting Wegner's theory.[3] If Wegner is right, or if a theory with the same general form as Wegner's is right, then the psychological system that generates the felt sense of effort is functionally distinct from the system that causes us to act. The system that generates the sense of effort leads us astray by causing us to feel ourselves the author of an action when we are not.

On Kane's view, there is a decisive act of mind, conscious and voluntary, in which the agent issues an imperative to herself. She declares that henceforth she is to become the kind of person who acts on this kind of motive. But again, what is the phenomenology of such acts? Even if Kane were right that we are the Romantic artists his theory describes, Wegner's epistemic defeater cuts to the bone. The theory of apparent mental causation shows that conscious and apparently voluntary acts such as these need not, and often do not, reflect the actual causes of our action. In this case, the actual causes of the agent's self-proclamation may appear to the agent to be precisely her conscious and voluntary endorsing of one motive over the other. She may sincerely and wholeheartedly believe that the imperative she gives herself is a product of the conscious and voluntary thought of opting for option A over option B, and she is probably dead wrong. Her conscious and voluntary opting for option

A—the thought she had just before deciding—may appear to her, with all the vividness of truth, to be the cause of her opting for A. Yet she is probably wrong.

The possibility of error in this sort of case is no mere logical possibility. We know we commit this sort of error. The evidence offered by Wegner cannot rationally be ignored without rigorous experiments demonstrating otherwise. The fact that Wegner's theory of apparent mental causation synthesizes laterally with Wilson's theory of causal interpretation and Ross and Ward's theory of naive realism only strengthens the point. So ask yourself whether you feel inclined to resist the conclusion I am drawing here, and if you do, ask why you feel this way. Is it because you have an alternative theory that synthesizes laterally with other well-confirmed theories and analyzes inward experimentally but that leads to a different conclusion? Does your inclination to resist my conclusion derive from considerations that ought to be granted authority in our inquiries concerning the self? At the very least, ask this: Kane says that he intends to put "the libertarian view into more meaningful dialogue with modern science." Has he done that? Can there be, at this point in the history of human inquiry, meaningful dialogue with science about the self if relevant theories in psychology and neuroscience are ignored?

It is precisely in light of these scientific theories that my directives come to the fore. Concepts dubious by psychological role—concepts such as 'reasons for acting' and 'ultimate responsibility'—must be divested of their former authority in the way we frame our inquiries. Moreover, the anticipatory systemic function of the capacities we are trying to study must take center stage. Our inquiries ought to be framed in terms of what we, by means of reliable methods, can glean about the anticipatory function of the capacities with which we generate reasons for acting or the capacities with which we assign praise and blame. Once we do that, our prior intuitions, hunches, or expectations regarding concepts dubious by psychological role are bound to lose their former grip. Such intuitions and hunches are relics of inquiry from a time when the study of our conscious experiences was beyond the reach of reliable scientific inquiry. The authority that Kane affords our conscious experiences must be withheld.

Conclusion

We are not the kind of agent that Kane takes us to be. This is no armchair speculation. It is something we know or, more modestly, something we believe with a preponderance of evidence. It is something we can deny only by turning our backs on the methods and findings of our best-developed sciences of the human self, the methods culled from the history of modern science, and lessons culled from the history of human culture. It should go without saying that our best-developed theories of today may be revised or refuted tomorrow; that is an intellectual instinct that the progressive orientation toward inquiry must cultivate. But it should also go without saying that the possibility of future theoretical advances should not stop us placing our best bets in light of what is currently known. We cannot, at any rate, bet on the basis of what is presently not known. We must settle for less; we must settle for what is known.

The central problem with Kane's libertarianism—the assumption that we are the conscious creators of our future selves—is hardly unique to Kane. In one form or another, it is also the assumption that defeats compatibilist views of free will, as we shall now see. Once we have the case against compatibilism on the table, it will become clear that the general thesis of this part of my discussion is difficult to deny.

NINE

Words Give Us a Special Ability: Compatibilist Conservatism

Words give us a special ability to deceive each other. There are many reasons to believe that animal behavior will lie to us less than human words. This dilemma is especially acute when it comes to our hidden feelings that we normally share only through complex personal and cultural display rules. **JAAK PANKSEPP, *AFFECTIVE NEUROSCIENCE***

We are not little Thomistic gods. The truth of evolutionary theory and recent developments in psychology and neuroscience should convince us of that. It may be possible to describe a version of Romantic libertarianism that is internally consistent, sitting prettily somewhere in logical space, a source of comfort and even gratification, but it would be no cause for celebration, since the largest regions of logical space are populated mostly by falsehoods. What matters to us—what should matter—is the balance of available evidence. It appears a fact of our history that, as we make progress in knowledge, more and more of our traditional view of the self is dying beneath the crushing weight of our knowledge. Like the growth of life, the growth of knowledge is fueled by death.

Yet it is precisely the death of the libertarian illusion that gives the alternative view of free will an appearance of reasonableness. It may even feel inevitable. If we are not Thomistic gods and if some concept or other of free will is to be saved, then our only salvation, it seems, is to retain whatever concept of free will we can consistent with a deterministic universe. That this is possible is the compatibilist's conceit. Like their libertarian cousins, compatibilists are committed to preserving elements of our experiences that appear constitutive of the kind of agent we are. But we should wonder. Today we know that the traditional opposition between science and experience is dissolving. So what do compatibilists see themselves as doing? Why the commitment to a view of freedom that is increasingly at odds with the growth of knowledge in psychology and neuroscience?

The answer, it seems, is that compatibilists are conservatives regarding our traditional notion of 'moral responsibility'; they aspire to retain at least some semblance of 'moral responsibility' as understood by our cultural ancestry. They do not aspire to a notion of responsibility as ultimate as Kane's, but neither do they aspire to a notion so anemic that we cannot make sense of our practices of bestowing moral and legal praise and blame. And precisely this aspiration dooms the compatibilist project from the outset. To convince you of this, I discuss a few representative theories in some detail and then generalize. I defend the claim that compatibilism is mistaken not so much in detail but because it adopts a view of human agency that recent progress in knowledge has called into question. Compatibilism is conservative and at times imperialistic regarding concepts dubious by psychological role and by descent. It is unresponsive to intellectual problems we now face because it remains transfixed by a problem that has passed away.

The Compatibilist's Motivation

Compatibilism does not assume that determinism is true. What inspires compatibilists is the apparently noble desire to preserve the concept 'moral responsibility' in such a way that it correctly applies to us even if determinism is true. Even if we believe that determinism is true, or even if we believe that it is impossible to know that determinism is false, still, we might aspire to demonstrate that human freedom and responsibility are nonetheless possible. In terms inherited from Samuel Johnson, compatibilists want to reconcile two things—our felt experiences and our scientific theories—that are locked in a mortal conflict. And they want to do this by showing that neither side needs to die.

It is worth emphasizing that compatibilists are realists about free will and moral responsibility, in this sense: to locate 'free will' and 'moral responsibility' is not to eliminate them but to show where in the natural world free will and moral responsibility genuinely arise. And this puts compatibilists in a delicate position. On the one hand, they eschew our theologically saturated notion of 'ultimate responsibility.' That, in fact, is probably why most compatibilists are compatibilists rather than libertarians—because they cannot square the existence of Thomistic gods and other metaphysical extravagances with the findings or methods of modern science. But neither, on the other hand, can compatibilists deflate too far the concepts 'free will' and 'moral responsibility,' since that would bring down charges of nonrealism regarding free will and responsibility.

Where then in the natural world should a realist look to locate, rather than eliminate, 'free will' and 'responsibility'? Indeed, the core question facing the compatibilist is, How do we reconcile the real existence of free will and moral responsibility with the existence of a universe that is, for all we know, deterministic? At least two general answers have been given.

The first asserts that our concepts 'free will' and 'moral responsibility' are located amid concepts and claims concerning the core deliberative capacities that make us rational practical agents. The centermost capacities we have for calm reflection, for weighing evidence pro and con, for imagining distinct routes into the future, for exerting our powers of attention, inference, and choice—all of these are at the heart of the compatibilist's location strategy. This is perhaps the most straightforward way to capture the intuitive pull of the compatibilist's conceit:

The control conceit: (1) The concepts 'free will' and 'moral responsibility' are correctly located in the domain of our core deliberative capacities, in choices or actions that result from the exercise of self-control, and (2) the human capacities for deliberation and self-control are not impugned by the truth of determinism.[1]

The intuition is compelling, at least on its face. Of all living things on earth, we are unique in our conscious capacities to anticipate the future, withhold certain things from ourselves or others in anticipation of an imagined future, cultivate certain things for ourselves or others, and so on. We appear unique in our capacities to exercise certain sorts of self-control with respect to our future choices and actions. Our capacities for control may well differ in kind from those of other primates, but, even if they do not, it appears just obvious that they differ in degree.

The second answer to the above question comprises two basic claims. The first is that 'free will' and 'moral responsibility' are properly located amid concepts and claims concerning the cognitive and affective capacities that cause us to develop deep social attachments. The claim is that attributions of moral responsibility have their home in forms of life underwritten by social emotions such as resentment and gratitude. This second claim is that the forms of life underwritten by our social emotions are effectively independent of any question concerning determinism. For organisms like us, who naturally feel resentment and gratitude toward one another and who, in consequence, naturally attribute moral blame and praise, the question whether determinism is true is simply irrelevant. This is a distinct form of the compatibilist's conceit:

The forms of life conceit: (3) The concepts 'free will' and 'moral responsibility' are correctly located in the domain of emotions and attitudes with which we form social attachments, and (4) the domain of social attachments constitutes a form of life that is partitioned, and should remain partitioned, from any general theoretical considerations (including the question of determinism).

The aim of the next two sections is to convince you that neither conceit is defensible. What defeats both forms of the compatibilist's conceit is the aspiration to preserve a view of human agency that progress in knowledge has called into question. In the final section of the chapter, I turn to Daniel Wegner's attempt in *The Illusion of Conscious Will* to preserve some semblance of our traditional notion of moral responsibility. I argue that, in light of the theory of apparent mental causation, his attempt does not fit well the progressive orientation toward inquiry. There is room even among our best naturalists to become better naturalists.

The Control Conceit: "Taking Responsibility"

John Martin Fischer and Mark Ravizza (1998) attempt to locate a concept of 'self-control' that is less than "ultimate" but fat enough to justify attributions of moral responsibility. The aim is to locate a concept they dub 'guidance control' among a set of conditions that obtain whenever human agents act freely. Two conditions form the heart of this account. An agent acts freely only when (A) the action is caused by the agent's reasons-responsive mechanisms and (B) the agent has, in the past, taken responsibility for her reasons-responsive mechanisms. The requirement in (B) is that a free agent must have "taken responsibility" for her reasons-responsive mechanisms. This, as we will see, is a particularly dubious requirement, but first a word about (A).

On the face of it, human beings have the capacity for deliberation and choice. How do we do it? By virtue of cognitive and affective mechanisms that make us responsive to reasons. Or so it is claimed. And these mechanisms fulfill a dual function. They make us receptive as well as reactive to reasons. Receptivity is the capacity to recognize the value of a possible action or state of affairs, where this recognition then functions as a motive in the agent's psychology, and reactivity is the capacity to choose, to produce a decision, on the basis of reasons recognized. Consider a person alive to aesthetic considerations. He values the power of opera to recreate features of the human condition in ways that elicit deep emotions. This makes him receptive to reasons of various sorts. It gives him a reason, by way of his receptive capacities, to spend large sums of money on opera tickets. And if our opera lover is also reactive to reasons, then under certain conditions his receptivity to reasons will cause him to deplete his bank account just for the love of opera.

It presumably makes no sense to say that an action belongs to an agent if it was not caused by capacities at her very core. This, at any rate, is an intuition at the heart of the control conceit. Moral responsibility seems to require that the agent's decision to act arise directly from her core deliberative powers. Her choices must be her own. They must emerge from where she, the practically rational agent, most fundamentally resides. Where is that? It is where she reflects, weighs, anticipates, focuses, infers, and chooses. The intuition then is that this deliberative core exists somewhere in or near the capacities with which the agent recognizes and responds to features of the world that engage her motives. This is the intuition motivating the requirement in (A).

The motivation for (B) is different. It is analogous to the intuition that drives Kane's concept of 'ultimate responsibility.' Kane is moved by the intuition that an agent is morally responsible for an action only if it was produced by elements in her character for which she is wholly responsible. Moral responsibility emerges in the process of self-creation. Fischer and Ravizza's appeal to "taking responsibility" is nothing more (but nothing less) than a surrogate for the libertarian's self-creation. There is no appeal to the Romantic ideal of overcoming adversity through the spontaneity inspired by struggle, but there persists a vestigial process in which the agent somehow "makes" certain parts of her psychology into "her own." What remains is the requirement that, in order to become a genuinely moral agent, persons must live through a historical process of "taking responsibility" for their deliberative core, for the psychological mechanisms that make them responsive to reasons. They must mature in the right way under the right sorts of conditions.[2]

This surrogate for Kane's "self-creation" is aptly described by Daniel Dennett as the process of "bootstrapping ourselves free." He says, "The last step required to complete my naturalistic account of free will and moral responsibility is to explain the [evolutionary development] that has given us each a perspective on ourselves, a place from which to *take* responsibility. The name for this Archimedean perch is the self" (Dennett 2003, 259). And Dennett tells us point blank that he is indeed doing what I say he is doing, namely, insisting upon a compatibilist surrogate to Kane's self-creation: "What, then, is the important role of such a self? The self is a system that is *given* responsibility, over time, so that it can reliably be there to *take* responsibility, so that there is somebody home to answer when questions of accountability arise. Kane and the others are right to look for a place where the buck stops. They have just been looking for the wrong sort of thing" (Dennett 2003, 287).

When he gets down to the business of telling us what the "self" is such that it "takes" responsibility, he says, "A proper human self is the largely unwitting creation of an interpersonal design process in which we encourage small children to become communicators and, in particular, to join our practice of asking for and giving reasons" (Dennett 2003, 273). Just so! Our concept 'moral responsibility' is located in the process of taking responsibility that occurs during social maturation. It occurs as developing agents are given responsibility by the full-fledged agents who oversee their development. It occurs, in particular, as young agents enter the social game of demanding reasons for the actions of others and giving reasons for actions of their own.

Fischer and Ravizza's view overlaps Dennett's, though they describe the process of "taking responsibility" in some detail. In the first stage, the child comes to see himself as a causally efficacious organism in the world. He comes to see that he can, by virtue of deliberating and forming intentions to act, inflict changes of various sorts on his environment. In the second stage, in addition to seeing himself as causally efficacious, the developing agent begins to see himself in normative terms. He comes to see that it is appropriate or "fair" to praise or blame his actions, at least under certain conditions. The six-year-old pilfers the cookie jar and, in response, we try to make him recognize that he must account for his action—we point out that no one was forcing him, that he could have asked permission or simply refrained, and so on. Once he sees this, he begins to see that it is appropriate for us to hold him to account for using his agential powers in the service of a thieving scheme. This is how adults begin to "give responsibility" to developing agents. Fischer and Ravizza describe the process in terms of shared, social norms: "The individual must see that in certain contexts it is 'fair,' in the sense of being part of our given social practices, for others to subject him to the reactive attitudes in certain circumstances. That is, he must see that it is an appropriate move in the relevant 'social game' to apply to him the reactive attitudes in some contexts" (Fischer and Ravizza 1998, 211). Taking responsibility is a matter of coming to see the rights and responsibilities that exist within one's social setting, of seeing that, like it or not, one's actions will be judged as good or as bad in accordance with the rules of one's "social game."

In the third stage, the child is in fact held to account by parents and other adults in the moral community. This completes the giving of responsibility and, presumably, the acquisition of the capacity to take responsibility. Parents express disappointment and impose sanctions, or express pride and offer rewards, with the expectation that the child will internalize this process and adopt the same attitudes toward himself: "When parents (and others) criticize and morally educate us (in part by treating us as if we are appropriate candidates for the reactive attitudes), they invite us to take this 'internal view' of ourselves. The appropriate correlation between internal and external reactive attitudes is one sort of indication that the individual addressed is a proper participant in the moral conversation" (Fischer and Ravizza 1998, 212). At this stage of development, the maturing agent has become a "proper participant in the moral conversation." This happens when he takes over the task, formerly filled by his parents, of evaluating his own actions in accordance with inherited social rules. To do this, he must make those norms his own. He must make them operative and efficacious within his own psychology. He must cultivate the internal disposition to judge his own actions in ways that correspond with the external norms by which others judge his actions.

These then are the basic steps by which an agent takes responsibility for his reasons-responsive mechanisms, though the occurrence of each step must be confirmed by relevant forms of evidence. The agent must come to see himself as an efficacious agent on the basis of empirical demonstrations. He must experience his intentions leading to actions that alter the world in some way. Moreover, he must have evidence, presumably from parents and teachers, that the responsibility he is being given—the responsibility he is being encouraged to take—is in some sense authentic. He must have evidence that the social game is real, that the social norms are binding, and that there is nothing capricious about the game. He must have evidence that, in entering into the norms of his culture, he is not submitting to the coercive whims of those who happen to be in power. Instead, he must have evidence that he is entering a social game that is authentic, at least in this sense: that by entering the game he is better able to live in accord with what he values most; that by taking the responsibility he is being offered, *his reason* for doing so is that his values are served by opting into the game.

And this should make us wonder. We should wonder whether the process of maturation, as described by Fischer and Ravizza, is sufficient for taking responsibility. It seems obvious that a developing agent cannot genuinely take responsibility for her reasons-responsive mechanisms unless the process is anchored in her character in some substantive way. Yet the process of maturation described above can surely proceed apace without being so anchored. Imagine a maturing agent who pretends to accept the norms of her community not in light of what she most values but because otherwise she would face harsh sanctions. Entering the social game enables her to protect one thing she values, her safety, at the cost of everything else she values. Has she authentically taken responsibility for her reasons-responsive mechanisms? Enough to justify attributions of moral praise or blame? More to the point, imagine an agent who as a matter of psychological fact has no genuine reason to adopt the rules of her community—her values would be better met by refusing such rules—but who falsely believes that her values are met by taking on those rules. Has she, the matured agent who emerges from the maturation process, authentically taken responsibility for her reasons-responsive mechanisms?

The process of taking responsibility is authentic, it seems, only if the maturing agent adopts the norms of her community in light of her own values and thus for her own reasons. Something along these lines is required by the control conceit. This is to take seriously the thought that the compatibilist's account of taking responsibility is indeed an approximation to the libertarian's account of self-creation. The sense in which an agent enters into the social game for her own reasons can be explicated in three steps.

- Taking Responsibility: The mature agent has taken responsibility for her reasons-responsive mechanisms only if, during the maturation process, she utilized *some* of her core deliberative capacities—*some* of her existing values and reasons—to endorse *the rest* of her core deliberative capacities and the actions that arise from those capacities.

It is the agent as she exists at present who must transform herself into a future self responsible for her future character and actions. It is not that she does it all by herself—parents and teachers must surely help—but at least one originating source of her future responsibility resides in her present deliberative capacities. If not, the compatibilist view fails the *excuse test* : an agent for whom Taking Responsibility does not hold can truthfully excuse her character and behavior by pointing out that all of her core deliberative capacities are effects of her genetic and cultural inheritance. She can excuse herself by pointing out that, as of yet, there is no sense to be made of the claim that her core deliberative capacities belong to her since she has not endorsed the indoctrination received at the hands of her elders. And if the agent refuses to make excuses for herself, then we, in order to say what is true, ought to point it out on her behalf. We should excuse her by noting that, if Taking Responsibility does not hold, then none of the agent's deliberative capacities genuinely qualifies as her own. This application of the excuse test shows that any form of compatibilism endorsing the control conceit, in order to get off the ground, must contain the theoretical resources with which to block such excuses.

In addition to Taking Responsibility, however, the worry about authenticity also requires that the agent take responsibility on the basis of reasons that genuinely belong to her and that are genuinely efficacious:

- Accuracy: The mature agent has taken responsibility for her reasons-responsive mechanisms only if, during the maturation process, she endorsed (the rest of) her core deliberative capacities on the basis of character-based reasons that are (1) derived from values she actually possesses and (2) among the actual causal factors in the process.

The reasons the agent employs in the process of taking responsibility—reasons that arise from some of her core deliberative capacities—must in fact belong to her character and be among the actual causes involved. Otherwise the agent fails the excuse test. If the reasons the agent gives are not reasons that belong to her character, or if they are not reasons that are in fact efficacious at the time, then the subsequent endorsement of her deliberative capacities will not have been accomplished by the relevant deliberative capacities of the agent herself. The act of endorsing her deliberative capacities will be, once again, an empty charade. The bootstrapping mechanism will fail to engage.

In order, therefore, to successfully take responsibility for one's core deliberative capacities, reasons one in fact possesses must be causally efficacious. However, in addition to successfully *taking* responsibility, there is the further issue of being *justified in attributing* responsibility. We are, of course, particularly interested in questions of attributing moral responsibility, since that is where our practices of giving and demanding reasons come into play. So something more than mere accuracy is required:

- Knowledge: In order for the agent to justifiably attribute moral responsibility to herself, she must know that the Accuracy assumption was met in the course of her maturation. In order for a third party to justifiably attribute moral responsibility to an agent, it must know that the Accuracy assumption was met in the course of the agent's maturation.

An agent is justified in holding herself morally responsible for her actions only if she knows that the stages in which she allegedly took responsibility were properly executed. She must know that responsibility was taken on the basis of reasons that actually belong to her character. Otherwise she has no justified grounds for attributing praise or blame to herself. A parallel requirement must be met by third parties who attribute responsibility. In the context of the control conceit, therefore, attributions of moral responsibility require considerable knowledge of the developmental details of the person to whom responsibility is attributed.

Of course, our ordinary practices suggest that we simply assume that Accuracy and Knowledge are satisfied. We tend to give ourselves and others the benefit of the doubt; we tend to accept that for most normal human agents the maturation process proceeds more or less as Fischer and Ravizza describe. But our usual tendencies in this regard do not justify our attributions of moral responsibility. What is typical need not be what is warranted. The requirements set forth in Accuracy and Knowledge cannot be dismissed simply because they appear out of the ordinary. The only way to defeat these demands is to demonstrate how the process of maturation reliably results in the authentic taking of responsibility in the absence of Accuracy and Knowledge. In the meantime, however, we must proceed on the assumption that our attributions of praise and blame are warranted only if the agent did in fact take responsibility for his deliberative capacities and only if we are in fact justified in believing the agent took responsibility.

The trouble with the control conceit then is that Accuracy and Knowledge conflict with what is known about our reason-giving capacities. Begin with Accuracy. We know that we sometimes believe or feel that we have acted from or for a specific reason when we have not. If Timothy Wilson (2002) is right that the human mind includes a system that produces causal interpretations of perceived events (including our own actions), and if that system produces causal hypotheses during or after the events occur, then the potential for interpretive error is enormous. Errors will occur when the nonconscious mechanisms that actually cause our actions do not match the hypotheses produced by our interpretive system during or after the action. The same point applies to our feelings of authorship. If Wegner (2002) is right that our minds contain a system that generates a feeling of having authored an action by virtue of a prior conscious thought, and if that system is sensitive only to a limited range of consciously accessible inputs, then once again there is considerable room for error. The nonconscious mechanisms that actually cause our actions may be quite different from what we consciously feel and believe to be the cause of our actions. The way things consciously feel regarding the causes of our actions can be woefully far from the truth.

Now consider Knowledge. Consider again Wilson's interpretive system. Although we do not know that we are always led astray by the outputs of this system, we do know that we are sometimes wrong in what we offer or accept as reasons for acting. The best we can do at present is point out that the reasons we give for our actions are more likely accurate when the situation is relatively simple, when our perception of immediate causal factors plausibly maps onto actual causal factors. And this means that, in many cases, we are faced with a powerful epistemic defeater: we cannot reliably tell from the inside whether the outputs of our interpretive system match the actual causes of our action. The same limitation holds in light of the theory of apparent mental causation. We do not know at present how to reliably identify cases in which the feeling of having authored an action reflects the actual causes of the action. It is precisely this ignorance that underwrites a similarly powerful epistemic defeater. Since we cannot from the inside reliably distinguish accuracy from error, we cannot, for any given action, claim to know that we (our conscious thoughts or intentions) were the salient causes.

Since neither Accuracy nor Knowledge is met, the process of "taking responsibility" described by Fischer and Ravizza fails to address the worry about authenticity. It is simply not plausible that 'moral responsibility' can be located in the maturation process that they describe. The crucial error occurs at the outset, in framing the compatibilist's conceit with a view of our agential capacities that conflicts with our best-developed theories of the self. The claim that we can, on the basis of our deliberative capacities, bootstrap ourselves into morally responsible deliberators is at odds with what we know about our selves. I conclude, therefore, that the control conceit is yet another failed attempt to employ the concept location strategy to a dubious concept. And this should convince us to withhold antecedent authority from 'free will' and 'moral responsibility' on the grounds that they are dubious by psychological role.[3]

The Forms of Life Conceit: Partitioning Social Attachments

I come now to the second conceit embraced by some compatibilists, comprising the twofold thesis that (1) 'moral responsibility' locates amid concepts and claims concerning our social emotions and (2) the domain of social emotions constitutes a form of life in which questions of determinism have no traction. This latter thesis is particularly striking. The claim is that the question whether 'moral responsibility' locates in a deterministic (or nondeterministic) universe is not a genuine question for agents like us. Ascriptions of moral responsibility are part of an autonomous form of life and perhaps a freestanding language game that the thesis of determinism cannot and should not penetrate.[4]

This, I will argue, is a view of audacious conservatism, even imperialism, that no progressive can abide. The thesis that 'moral responsibility' locates amid our reactive attitudes flouts several of my directives for inquiry, as does the thesis that our social attachments constitute a form of life partitioned from all general theoretical questions. The first thesis fails to analyze inward to what is known about the neural bases of social emotions, flouts the directive in (DD) to withhold antecedent authority to concepts dubious by descent, and very likely flouts the directive in (DP) to withhold antecedent authority to concepts dubious by psychological role. The second thesis similarly ignores several directives. Perhaps most egregious is that it deliberately refuses one of the most basic mechanisms of self-correction in scientific inquiry, the directive to synthesize laterally with what is known in other domains, and then tries to make an intellectual virtue of it. These are strategies of philosophical reflection toward which we should have no tolerance. Let us, at any rate, have a look.

Locating 'Moral Responsibility'

I begin with the first thesis of the forms of life conceit. According to its best-known advocate, Peter Strawson, we tend to feel certain affective attitudes in response to perceived social offenses. This, he claims, is a natural feature of our psychology. Affective attitudes allegedly fall into three distinct categories, but it suffices to focus on one such category, namely, personal attitudes. These are natural reactive attitudes we take toward others in response to their attitudes or behaviors toward us. We feel resentment, for instance, when we perceive that someone is indifferent or has injured us. If I perceive that you have insulted me publicly, my feeling of resentment is immediate and viscerally negative; it is not the conclusion of argument but the simple and unvarnished expression of my psychology. My resentment, however, can be diminished or dissolved under some circumstances. If I discover that you are being terrorized by someone else and that your well-being is at risk if you fail to publicly insult me, my natural resentment will diminish. This too is an immediate, visceral, unvarnished expression of my psychology. It is, more generally, an instance of what Hume called "the natural sentiments of the human mind." Resentment and the other reactive attitudes are among the naturally entrenched social sentiments of the human animal.[5]

It is crucial that, according to Strawson, feelings of resentment for perceived injuries involve a *demand* that we make on one another:

We should think of the many different kinds of relationships which we can have with other people .... Then we should think ... of the kind of importance we attach to the attitudes and intentions towards us[,] of those who stand in these relationships to us, and of the kinds of *reactive* attitudes and feelings to which we ourselves are prone. In general, we *demand* some degree of goodwill or regard on the part of those who stand in these relationships to us, though the forms we *require* it to take vary widely in different connections. (Strawson 1982, 63–64; last two italics mine)

What causes and perhaps justifies our resentful reaction is the implicit demand that the offending person not behave as she has done. My resentment toward you is caused and perhaps justified by the apparent requirement that you not behave in that way and the perception that you did indeed violate this requirement. I will return to this alleged imposition of demands presently.

Two further features of reactive attitudes deserve our attention. The first is that they are not the only natural attitudes we experience in social interactions. We also experience attitudes of detachment. When I learn that someone is threatening you lest you harm me, my resentment toward you is tempered and perhaps eliminated. Why? Because I cease seeing you as a full-fledged agent in this circumstance and instead come to see you as a hostage in need of rescue or, more commonly, a patient in need of rehab. In consequence, I suspend what Strawson calls the "participant" attitude toward you and instead take an "objective" attitude. And the disposition to take the objective attitude is yet another natural feature of our psychology.

And yet, though the objective attitude is something that comes naturally, it is not constitutive of social relations and cannot be sustained for any considerable length of time. Or so it is claimed. This is the second feature that applies to all reactive attitudes, a feature thought to operate at our psychological core:

The human commitment to participation in ordinary interpersonal relationships is, I think, too thoroughgoing and deeply rooted for us to take seriously the thought that a general theoretical conviction [for example, that determinism is true] might so change our world that, in it, there were no longer any such things as inter-personal relationships as we normally understand them; and being involved in inter-personal relationships as we normally understand them precisely is being exposed to the range of reactive attitudes and feelings that is in question. (Strawson 1982, 68)

If we perversely adopt the objective attitude to everyone we meet for any length of time, we will undermine all our social relationships. We will cease treating others as persons or as agents endowed with capacities of self-control. And that appears to be a course of action that no human being can sustain. The nature of human interpersonal relationships "as we normally understand them" is so deeply embedded in the fabric of our psychology that nothing, not even the fruits of inquiry, can dislodge them.

How then does 'moral responsibility' locate amid these concepts and claims? Consider again the situation in which I naturally experience resentment toward you. The suggestion, I take it, is that in this situation I also naturally attribute moral blame to you for insulting me. And my attribution is justified in so far as the experience of resentment includes an implicit demand or requirement that you, by insulting me, failed to meet. Something similar holds in the relationship between gratitude and moral praise. I feel gratitude toward you for rescuing my purse from the clutches of a thief, and it is natural for me and others to attribute moral praise to you. The attribution is justified because the experience of gratitude includes an implicit demand that you satisfied (and probably exceeded) with effort and bravery.[6] There is, it seems, something naturally binding about the reactive attitudes, and it is, it seems, precisely these natural bonds that underwrite and confer authority on our attributions of moral praise or blame. They appear to be not merely the causal but also the legitimating source of our attributions. This then is the first thesis in the forms of life conceit.

But so long as we bear in mind two simple facts, this first thesis seems an obvious failure. The first fact concerns the compatibilist's conceit generally. The conceit, recall, is that we can reconcile the real existence of free will and moral responsibility with the existence of a universe that is, for all we know, deterministic. The conceit is that we can preserve a notion of free will robust enough to justify our attributions of moral and legal praise and blame. And, as I have said, that puts the compatibilist in the delicate position of trying to save rather than eliminate some version of our traditional notions of freedom and moral responsibility without endorsing anything so theologically saturated as libertarian freedom. The second fact concerns Strawson exegesis. His discussion of freedom and resentment is framed in terms of the disagreements between libertarians and compatibilists. Although his sympathies clearly tend toward compatibilism, he tries to take seriously the libertarian's complaint that compatibilists are facile in their willingness to relinquish the robust notion of freedom that runs deep in our cultural heritage and that libertarians wish to save. Strawson's aim, that is, is to give libertarianism its due, or at least some of its due, even while defending compatibilism. And this aim dictates his strategy. If it can be shown that our belief in free will and our attributions of moral responsibility are anchored in the very constitution of our psychology, in capacities we cannot and would not change or relinquish, that would show that neither our freedom nor our attributions of moral praise and blame are as ephemeral or shallow as they might otherwise appear. It would show that there is something deep and secure about free will and moral responsibility as conceptualized by the compatibilist. That, perhaps, would allay at least some libertarian anxieties.

If, therefore, Strawson is a realist compatibilist about 'moral responsibility,' and if, in particular, he wishes to draw the libertarian to his side, then the notion of 'moral responsibility' relevant to his discussion cannot be anemic. He must be endeavoring to locate a concept of moral responsibility that is sufficiently fat in its normative dimensions; he must be trying to save a version of the concept in which at least some of the uniquely moral demands are preserved. Strawson, in short, is best construed as offering a genuine form of location, rather than elimination, regarding some version of our traditional notion of moral responsibility. He does not tell us with any precision what notion he is talking about, nor is it crucial that we settle on any precise rendering of the concept.[7] What matters is that, in light of the two facts just described, Strawson's proposal cannot succeed unless some of the normative dimensions of our traditional concept are traced to the natural demands involved in our reactive attitudes.[8]

And on this construal, Strawson's concept location project is an unambiguous failure. The question to ask of his proposal is whether the proposed mechanisms or processes do indeed constitute free will and moral responsibility. The question is whether the attempt to analyze inward from 'moral responsibility' to 'resentment,' 'gratitude,' and the rest leads us to reasonable evidence that these reactive attitudes do indeed give rise to demands that are not merely social but distinctively moral as well. The answer, of course, is that no such evidence has been given.[9] Strawson gestures in the direction of an analysis. He appeals to intuitions regarding various cases and insists that we are driven by the natural bonds of our reactive attitudes to see our relationships with one another in terms of moral bonds. But of course all that may be true in a world in which moral responsibility does not exist at all. The ways in which we are caused by our psychology to see or feel our social relations is not a reliable guide to the existence of any moral bonds.

In fact, the gap between our traditional notion of moral responsibility and the capacities with which we form and sustain social attachments is much wider than Strawson lets on. This becomes clear once we make even a modest attempt to analyze inward. Consider the framework developed by Jaak Panksepp comprising two suggestive hypotheses. The first concerns the sources of our social attachments within our psychology and neurology: "The most reasonable supposition at present is that social bonding ultimately involves the ability of young organisms to experience separation distress when isolated from social support systems and to experience neurochemically mediated comfort when social contacts are reestablished" (Panksepp 1998, 274). The suggestion is that we are motivated by the structure of our affective capacities to develop social attachments as a response to separation distress. The experience of separation or loss, or at least the perceived threat of separation or loss, causes in us a powerful form of stress that motivates us to develop and sustain social attachments. On this view, we are not possessed of a neural system that independently moves us to form social attachments; we do not have a system dedicated to social attachments in the way, for example, our digestive system is dedicated to digestion. Instead, the tendency to develop social attachments is a reaction to the crushing stress of grief or loneliness and perhaps also a way of anticipating such stress. It is a compensatory attempt to restore or maintain emotional equilibrium.

The available evidence, at any rate, is that our natural orientation to develop social attachments serves two systemic functions. It quells feelings of distress that come with separation and restores us to equilibrium. It also serves the anticipatory systemic function of keeping such feelings at bay and thus maintaining equilibrium. Neither function is trivial. The distress experienced by infants and very young children when separated from others, especially their mothers, is the overwhelming terror of panic that runs deep in our neural constitution and generates physiological responses akin to those we suffer when we cannot catch a breath.

Evidence for this hypothesis is striking once we begin to analyze inward.[10] When young rats are socially isolated, several indicators of stress manifest themselves. They emit distress calls for a short period of time, but they also endure longer lasting changes such as a decrease in body temperature, disruptions in sleep, an increase in the production of the stress hormone corticosterone, an increase in overall behavioral activity and sucking tendencies, and so on. When social contact is reestablished and especially when there is friendly or playful touching, the distress calls cease, the body warms, the production of corticosterone falls off, and the body produces more opioids and oxytocin. The pain of distress is replaced by the pleasure of comfort. We know, moreover, that the same effects can be produced artificially by altering the neurochemistry of the organism. Injecting chemicals that inhibit the production of endogenous opioids or oxytocin causes an organism to cry even when its social contacts are intact. And injections of opioids or oxytocin shut off an organism's distress cries even when it is separated from family and friends.

Panksepp's second general hypothesis concerns the way in which our disposition toward social attachments likely evolved. This is to place his first hypothesis in a plausible evolutionary framework. Our natural disposition to form and perpetuate social attachments probably evolved as a spin-off from what Panksepp calls the PANIC system in the brain. This is a system of the mammalian brain comprising various structural and neurochemical interactions that generate the distress caused by separation and loss:

The PANIC system: To be a mammal is to be born socially dependent. Brain evolution has provided safeguards to assure that parents (usually the mother) take care of the offspring, and the offspring have powerful emotional systems to indicate that they are in need of care (as reflected in crying or, as scientists prefer to say, *separation calls).* The nature of these distress systems in the brains of caretakers and those they care for has only recently been clarified; they provide a neural substrate for understanding many other social emotional processes. (Panksepp 1998, 54)

Our brains appear to include mechanisms that govern our affective responses to the actual or potential experience of separation and loss. Moreover, the evolution of this disposition is a likely case of coevolution. The disposition of young children to cry out in distress when separated from their mothers probably evolved hand in hand with the disposition of parents to madly search out their children when such cries reach their ears: "Exactly how a concerned attitude is promoted within caretakers' brains by hearing distress calls from their infants remains unknown, but I would suggest that the sounds of crying arouse distress circuits in parents that parallel those of the children. If so, associated learning systems may rapidly establish the knowledge that an optimal way to reduce distress in both is for the parents to provide care and attention to their offspring" (Panksepp 1998, 266–67). Though Panksepp is careful to say that this is conjecture, he has the accumulated weight of parental experience on his side. The sound of my daughter's cries at night, especially when she was a baby or toddler, would have me out of bed and on my feet, halfway to her room in the grip of a powerful adrenaline surge, all before becoming conscious of where I was.

It is plausible, moreover, to conceptualize the distress of separation and loss as, quite literally, a form of pain. This is supported by behavior and neural anatomy. Crying in response to pain is something we do long before our social capacities mature. Yet when those capacities become operative—when young mammals begin to respond behaviorally to separation—they do so by crying, by adopting a behavior already in their repertoire. The neural mechanisms that produce separation distress, moreover, overlap the more ancient mechanisms involved in physical pain. This is so, at any rate, in primates, cats, chicken, and guinea pigs, in which, as Panksepp puts it, "the PANIC system appears to arise from the midbrain PAG, very close to where one can generate physical pain responses. Anatomically, it almost seems that separation [distress] has emerged from more basic pain systems during brain evolution" (Panksepp

1998, 267). So the distress in young mammals caused by separation probably evolved out of more primitive neural mechanisms that produce physical pain.

Panksepp also offers the following suggestive remark about the interactions between children distressed by separation and parents distressed by the distress of their children. He says, "In any event, when care is provided, emotionally distressed children rapidly exhibit responses of comfort and satisfaction, even though, if the care has taken too long to arrive, they may also *harbor some resentments,* as indicated by a transient phase of *social detachment upon reunion.* Adults often do the same. Apparently, through such social reciprocities, the social bond between related animals is first established and periodically strengthened" (Panksepp 1998, 267; my italics). The suggestion is that resentment—or something we may plausibly call "resentment"—is a natural reaction to a disappointed, implicit expectation regarding pain. If the timing is off and the distressed child suffers too long—either the pain of separation or the pain of the threat of separation—the subsiding symptoms of distress are replaced by detachment. The child refuses, if only for a short while, to accept the offer to reestablish contact.

How then should we conceptualize the dispositions we have for resentment and gratitude? Hypotheses along the following appear to capture the facts just described. Some social emotions are negatively charged affective states caused by disappointed needs or expectations regarding the pain of separation. When the expectation of relief from pain is disappointed, feelings of anger or hurt may ensue, resulting in the transient refusal to reestablish contact. Such feelings we typically call "resentment." In addition, some social emotions are positively charged affective states caused by the satisfaction of needs or expectations regarding the avoidance of separation or the pleasure of social contact. What we typically call "gratitude" is plausibly a form of pleasure caused when the distress of separation is alleviated or when the threat of separation is removed or, more simply, when the pleasures of social contact occur and stave off the possibility of separation. Conceptualized in this way, our social emotions should be understood first and foremost as the affective effects of our expectations regarding the behavior of others in the production, elimination, or prevention of pain involving separation. On this view, the "demands" of our reactive attitudes to which Strawson alludes are natural expectations we have toward one another regarding the pains and pleasures of social interactions.

Cast along these lines, it is abundantly clear that the reactive attitudes cannot be the source of genuinely moral demands. The gap between them is too great. As we have seen, Strawson's proposal seems to be that the implicit demands of our reactive attitudes somehow constitute and legitimize the demands of moral responsibility. And yet when you publicly insult me, you disappoint my natural expectation that you will refrain from inflicting on me the threat of social isolation that a well-placed insult can cause. My natural reaction is to feel resentment toward you. But my disappointed expectation explains why I feel negatively and detach behaviorally from you. It does not explain how or why I am justified in *morally* blaming you. That you have thwarted my expectations may cause me to raise my voice and speak to you in the imperative mood. It may cause me to demand that you not cause me such pain ever again. It may even prompt me or some segment of the community to impose its will and punish you for what you have done; we may insist that you be "held to account." But the force of the demand provides no traction for moral responsibility as traditionally conceptualized. Strawson flouts the expectation of change in our conceptual categories as we analyze into the relevant natural systems.

He also flouts the directives concerning dubious concepts. That 'moral responsibility' as traditionally understood is dubious by descent is difficult to dispute. But it is also dubious by psychological role. This is suggested, in fact, by Strawson's own view! He seems to be suggesting that we are caused by our reactive attitudes, by the bonds imposed by our social nature, to see ourselves and others as free and as subject to the bonds of morality. The suggestion is that when we experience the felt effects of our social bonds, our feelings prompt us to see or feel those bonds not merely as social but as something different. One part of our psychology causes another part to conceptualize our relations to one another in terms that are distinctively moral. And this is to assume that we are indeed subject to a constitutional conflict regarding attributions of moral responsibility. It makes no difference whether our traditional concept of moral responsibility is empty. If it is empty, then the conflict is that we are led by our reactive attitudes to see our relations to one another in terms of something that does not exist. But even if the concept is not empty, even if our traditional notion refers to something altogether real, then the conflict is that we are led by our social attachments to falsely see our reactive attitudes to one another as sufficient to implement the bonds of morality—that is what the failure to analyze inward shows. So Strawson's own account of our reactive attitudes, in so far as it is plausible, should compel us to the conclusion that 'moral responsibility' is indeed dubious by psychological role, since the capacities that cause us to apply that concept cause us to accept as true something that is false.

The first thesis of the forms of life conceit, therefore, fails to adhere to even the most elementary directives for inquiry. It fails to analyze inward; one of the lessons that motivate the expectation of conceptual change in (EC) is ignored. And the directives in (DD) and (DP) concerning dubious concepts are likewise ignored. Finally, the more specific directive in (CL) concerning the concept location project in general is also ignored. The effects of ignoring these directives should not be ignored. Any attempt to locate a concept as dubious as our traditional notions of free will or moral responsibility amid the concepts and claims of any well-developed science is, on its face, not a serious form of inquiry or reflection. It flies in the face of the progressive orientation and the destructive nature of serious scientific inquiry.[11]

Partitioning 'Moral Responsibility'

The problems with the second thesis of the forms of life conceit are even worse. Imagine a critic of Strawson's view who grants the central descriptive claim that we tend to attribute moral responsibility when we feel certain reactive attitudes but who queries whether doing so is rational. In raising this question, the critic makes two assumptions. He assumes that our reactive attitudes toward one another do not exhaust the morally binding demands that apply to us all. This seems reasonable enough, since we seem able to imagine cases in which our reactive attitudes prompt us to do what we judge to be immoral. The critic also assumes that some of our natural dispositions can be altered. Here too he seems to have a point, since we often check our reactive attitudes or our behaviors in an effort to do what is morally right. We often check ourselves, that is, in the hope (and expectation) that our attitudes can be altered by force of habit. So the critic's question must be addressed. That is, even if we are naturally disposed by our reactive attitudes to attribute moral praise or blame to one another, why think our attributions are justified? Why think that what is natural suffices for what is moral?

Strawson anticipates this question and gives an entrenched reply. Referring to what he calls our "natural human commitment to ordinary inter-personal attitudes," he says, "This commitment [to ordinary interpersonal attitudes] is part of the general framework of human life, not something that can come up for review as particular cases can come up for review within this general framework. And ... if we could imagine what we cannot have, viz. a choice in this matter, then we could chose rationally only in the light of an assessment of the gains and losses to human life, its enrichment or impoverishment; and the truth or falsity of a general thesis of determinism would not bear on the rationality of *this* choice" (Strawson 1982, 70).

This is an extraordinary claim. It is impossible, because contrary to our nature, to change our reactive attitudes: "it is *useless* to ask whether it would not be rational for us to do what it is not in our nature to (be able to) do" (Strawson 1982, 74; Strawson's italics). And even if the first point were false—even if we could alter our nature in this regard—it would be irrational to jettison the reactive attitudes, because life without reactive attitudes and, in consequence, without interpersonal relationships would be grossly inferior to life as we know it. This then is the second thesis of the forms of life conceit.

But this second thesis is even less plausible than the first.[12] It is hard to imagine a more conservative approach to the study of human agency. Instead of adducing evidence that our reactive attitudes are immune to change, Strawson gestures toward thought experiments in which all our reactive attitudes go missing from our social interactions. Suddenly no one takes the participant stance toward anyone else, everyone sees and approaches everyone else as a patient or as something subagential. In this vague, nightmarish scene, we are told that all our social attachments would be drained of their blood. We would be reduced to the cold calculations of the objective stance, bereft of love, gratitude, intimacy, trust. The imagined form of life is indeed a nightmare, but—lucky for us—it is also a form of life is so alien that we cannot see how to get there from here. And that, presumably, is what makes it rational to think that changing our reactive attitudes is indeed impossible.

We should not fall for this line of thought. It takes a remarkable failure of imagination to convince oneself that our reactive attitudes are constitutionally resistant to change. Indeed, for any of the capacities we have as agents, the thesis that they cannot change is flatly implausible in light of what we know about our actual history and constitution. But, apart from the facts of our evolutionary history, the very strategy with which

Strawson argues belies a failure of imagination. It is unlikely, after all, that members of our species would suffer a sudden loss of social emotions and attachments. Changes to our affective and cognitive capacities, if they were to come, would probably not be the products of dramatic surgical strikes.[13] They would more likely be gradual alterations that trigger myriad changes elsewhere in our psychology. Some of these changes might be slow to emerge, others might be precipitous; some might undermine our capacities as agents, others might enhance them. The bottom line, however, is that we presently have *no idea* what sorts of changes we would suffer as a result of losing our reactive attitudes. We simply do not have the requisite knowledge to rationally speculate on the historical trajectory that this would involve. So the alleged "experiment" in thought simply cannot be completed. We are asked to delete in our imagination all our affective attitudes, but we find ourselves utterly clueless of what to put in their place. It is then suggested that the resulting gray and depressing picture is the obvious outcome of the experiment. But it is not the outcome of the experiment, or, more precisely, it is not an *experimental* outcome. It is merely the unhappy effect of an *unfinished* experiment. It is incomplete because our ignorance stultifies our imagination; we know too little to complete the thought experiment.

Indeed, the analogy to Darwin's reasoning at this point should not be missed. He saw that the reasoning required to appreciate the power of his theory demands more than our undeveloped imaginations can deliver. He saw that the only way to grasp the power and scope of his theory is to resist the usual limitations on our historical imagination by training ourselves to see all of life in deeply historical terms and to resist the usual limitations on our psychological imagination by learning to see, for example, the incessant violence and destruction required for life. By analogy, we should suspect that the only way to discover what is possible regarding the plasticity of our current capacities and the evolution of different capacities is to ignore what our imaginations seem to suggest and insist on studying the actual mechanisms involved. We must train ourselves to resist the temptation to think that the way things appear to us in our imaginations corresponds to the way things in fact are or the various ways things could be.

Strawson also asserts that even if it were possible to change our reactive attitudes, it would be irrational to do so. But it should be clear by now that, because Strawson is in no position to describe what our lives would be like without the reactive attitudes, he is in no position to evaluate the claim that our lives would be worse without them. And there is also this: when we try to guess the sorts of changes in reactive attitudes that we might survive, it appears just as reasonable to speculate that our lives would not be diminished by the loss of our reactive attitudes. Why? Since we have no idea what life without the reactive attitudes would be like—since our imaginations lack the information required to complete the thought experiment—none of the possible outcomes can be ranked above the others. For all we can determine, life without the attitudes would be progress. Additionally, and a bit more substantively, it appears evident that we humans are natural-born generators of meaning. Our psychology appears geared toward the future, toward what is next, and thus our response to the loss of the attitudes may be to orient ourselves toward the future in some other way. If our capacities evolve in ways that eviscerate the drive to find meaning in social attachments, but we nonetheless remain driven to find meaning, then we will no doubt try something else. We might fail, but we might not. This speculation, at any rate, is no more far-flung than Strawson's.

The most decisive problem, however, is not the failure of imagination but the more devastating point that the second thesis absurdly turns its back on an elementary directive of inquiry, namely, that we make progress in knowledge as we synthesize laterally. To turn our back on this directive is an especially retarding effect of the default assumption of all theology, the assumption that we already know, prior to inquiry, what we or our capacities as agents are actually like. It is precisely here, in claiming that our reactive attitudes are and must remain partitioned from all other areas of inquiry, that Strawson's conservatism spills over to imperialism. After all, lateral synthesis presumes prior success in analyzing inward, in identifying low-level mechanisms and relations that implement high-level systemic capacities. Once we have what appears an adequate taxonomy of low-level mechanisms and relations, we are directed to relate our taxonomy to well-confirmed theories and taxonomies in other areas. We are directed in this way for the compelling reason that a convergence of taxonomies across a wide range of related theories is a *reliable indicator of the truth* of one's theory. Conversely, the failure to synthesize laterally is a powerful check on any theory. So long as we are committed to seeking the truth, the directive to analyze inward and synthesize laterally applies.

It is, as I say, absurd to adopt an orientation that refuses to employ a strategy as powerful as this. It is absurd in Strawson's case to insist that human nature is incapable of evolving in ways that fundamentally alter the reactive attitudes; it is absurd to insist that our entrenched dispositions to attribute moral praise and blame are rightfully partitioned off from all general theoretical considerations. Such insistence is absurd because it amounts to the stubborn but uninformed—indeed, unscientific—refusal to synthesize across other domains of inquiry. An orientation toward inquiry that diminishes in this way our chances of discovering the truth deserves our impatience, even our contempt.

The Defeater's Defeat

The conclusion so far is this: the compatibilist's conceit in both forms is defeated by the directives for inquiry. It is defeated by resisting the conservatism and imperialism of philosophers and theologians and by employing the discoveries and progressive orientation of scientists. We should not think, however, that even our best scientists are altogether immune to the retarding effects of conceptual conservatism. When it comes to concepts with the greatest apparent importance—when it comes to our concept 'moral responsibility,' for instance—we should expect to encounter lingering reluctance to cut ourselves so decisively adrift. That is indeed what we encounter. I conclude this chapter by considering one such instance.

As we saw in chapter 7, Daniel Wegner approaches the question of free will scientifically by focusing on the felt experience of consciously willing an action, precisely the sort of experience Johnson refers to in his famous aphorism. But Wegner does not naively take such experiences at face value. Instead he asks, When we analyze inward from the felt experience of consciously willing an action to the low-level mechanisms that cause us to act, and when we synthesize laterally with other, well-developed theories in psychology and neuroscience, what do we find? The answer—or one plausible answer—is the theory of apparent mental causation. The answer is that we have a psychological system that generates in us the felt experience of having willed our own actions, a system triggered not by the myriad nonconscious causes of our actions but rather by a limited range of consciously accessible experiences. We are, on this theory, constitutionally conflicted with respect to the causes of our own actions. We thus must withhold, in light of the directive in (DP), antecedent authority to our concept of 'free will' and to other associated concepts.

The power and momentum of Wegner's argument thus drive us to the edge of the precipice, forcing a cluster of traditional concepts to the very brink. But, at the last moment, Wegner pulls us back. The question Wegner poses for his final chapter is this: "Why, if this experience of will is not the cause of action, would we even go to the trouble of having it? What good is an epiphenomenon? The answer becomes apparent when we appreciate conscious will as a feeling that organizes and informs our understanding of our own agency. Conscious will is a signal with many of the qualities of an emotion, one that reverberates through the mind and body to indicate when we sense having authored an action" (Wegner 2002, 318). In the same paragraph, Wegner points out that his answer to the question is novel. He says, "The idea that conscious will is an *emotion of authorship* moves beyond the standard way in which people have been thinking about free will and determinism, and presses toward a useful new perspective. This chapter explores how the emotion of authorship serves key functions in the domains of achievement and morality" (Wegner 2002, 318).

Now, to the best of my knowledge, Wegner's theory of conscious will is indeed new, but does it nevertheless try to conserve what ought to be discarded? Does it, in particular, try to save what it has already called into question?[14]

The thesis of Wegner's final chapter is that, although the felt experience of will is illusory, this feeling nevertheless serves the important functions of, first, reliably indicating to the agent when she is the author of an action and when she is not and, second, signaling to the agent actions for which it is appropriate to feel moral emotions and attribute moral responsibility. The thesis is that there is an important concept of moral responsibility that can be saved despite the theory of apparent mental causation. I begin with the discussion concerning achievement and work my way toward responsibility.

The claim is that the feeling of willing is a reliable guide to what we have indeed accomplished. This is not to say that the feeling of willing is always correct. Often our perceptions of control do not match the causal facts involved, as the literature on "perceived control" suggests. Still, the feeling of conscious willing is accurate enough to indicate to us which actions we have authored and which we have not. Of course, our feeling fulfills this function only to the extent that it is reliably correlated with the actual causes of our actions, with the empirical will. In a remarkable passage, Wegner puts it this way:

Conscious will is the somatic marker of personal authorship, an emotion that authenticates the action's owner as the self.... Often, this marker is quite correct. In many cases, we have intentions that preview our actions, and we draw causal inferences linking our thoughts and actions in ways that track quite well our own psychological processes. *Our experiences of will,* in other words, *often do correspond correctly with the empirical will—the actual causal connection between our thought and action.* The experience of will then serves to mark in the moment and in memory the actions that have been singled out in this way. We know them as ours, as authored by us, because we have felt ourselves doing them. (Wegner 2002, 327; my italics)

The story then is this: often the feeling of willing correctly *corresponds with* the actual low-level causes of our action and, by virtue of this correspondence, the feeling of willing reliably indicates the extent of our control or achievement, while the absence of this feeling reliably indicates that we have hit upon something real that we cannot control.

I do not think we can accept the claim concerning correspondence—not, at any rate, if we accept the theory of apparent mental causation. Nor can we accept that the feeling of willing serves the functions that Wegner says it serves. I will focus my criticisms on the correspondence claim. I will show that, if we accept the theory of apparent mental causation, we thereby accept that the human mind is *constitutionally conflicted* with respect to the causes of our actions. The conflict produced by the structure of our psychology is that we cannot pretend to know, from the first-person perspective, the causes of the actions we perform. The experiences of unfettered and self-directed actions—the experiences that persuade us that we are free—are prone to lead us astray, in which case the directive in (DP) to withhold antecedent authority to concepts dubious by psychological role must be applied. In defense of this claim, I will consider three specific points.

To see the first point, consider the following levels of processing in Wegner's model:

1. the feeling of willing (arises from)
2. drawing a causal inference between conscious intention and perceived action (caused by)
3. the perception of priority, consistency, and exclusivity between conscious intention and perceived action

Now the correspondence claim asserts that there is a positive correlation—a "correct correspondence"—between level 1 and a further level of processing, namely,

4. the empirical will—nonconscious, low-level causes of action

But this claim of correspondence, in the context of Wegner's larger view, is implausible on its face.

Wegner has provided a theory concerning the relations between levels 1, 2, and 3. If his development of Michotte's theory concerning the perception of causality is correct, then correlations between 1, 2, and 3 are nonaccidental. They hold because of our psychological architecture. They hold because an interpretive system responsive to the perception of priority, consistency, and exclusivity among pairs of events generates a causal inference, which in turn generates the cognitive emotion of authorship. None of this, however, can be said about the relation between levels 1 and 4. We have no psychological theory—no theory concerning low-level architecture—linking the conscious feeling of willing to the nonconscious, empirical will. In fact, to the contrary! We have a wealth of considerations suggesting that, within the architecture of the human mind, these two processes—the feeling of will and the empirical will—run on more or less independent tracks. This is the take-home lesson of chapters 4, 5, and 6 of Wegner's book. The suggestion at the close of chapter 4, in light of the ideomotor theory of action and other considerations, is that we should conceptualize our psychology as consisting *fundamentally* of automatic, nonconscious processes, with the conscious will being an add-on in need of special explanation. Automatisms are the rule; consciousness, the exception. The suggestion in Wegner's chapter 5 is that our interpretive system operates with a template of "the ideal causal agent" and that this system works incessantly to forget or fabricate conscious intentions in order to conceal the extent to which our agency is nonideal—hence the efficacy of mechanisms such as cognitive dissonance. And Wegner's suggestion in chapter 6 is that our interpretive system, under a range of conditions, undermines the emotion of authorship even when our nonconscious processes are clearly the cause of our action. The unfortunate case of facilitated communication vividly illustrates this suggestion, as do the experiments reported by Wegner, Fuller, and Sparrow (2003). So we cannot accept the theory of apparent mental causation and also accept that the feeling of willing corresponds with the empirical will.

My second point is that even if there were some sort of correspondence between levels 1 and 4—even if my first point could be defeated—we nevertheless have good grounds for denying that the correspondence is ever "correct." Three brief considerations support this point.

1. Recall that, on the theory of apparent mental causation, the causal relata are consciously accessible intentions and perceptions of oneself acting. So, in any given case, the inference drawn must be something like

(CI) My consciously accessible intention to do A caused my doing A.

The subsequent emotion of authorship, therefore, should be something akin to

(CF) The feeling that *I,* by virtue of conscious, intentional effort, caused my doing A.

That is, the feeling of willing is a feeling of consciously exerting the effort of initiating and executing A. But, of course, it is false that what occurred in consciousness caused A; apparent mental causation is, after all, apparent causation. The emotion of authorship mistakenly leads us to specify the cause of our action in consciousness, while the actual cause operates beneath the level of conscious awareness in the empirical will. So we have two quite different sorts of objects, a conscious intention that does not cause our action and a nonconscious process that does cause our action. In what sense then does the former "correctly" correspond to the latter? Even if we grant that a correspondence of some kind obtains, on what grounds should we agree that the correspondence is "correct" as opposed to "incorrect" or as opposed to "neither correct nor incorrect"?

2. The second consideration goes further. The problem is not simply that the content of the conscious feeling of willing is in error. Suppose we assume that nonconscious processes that cause us to act—those operating in the empirical will—also have intentional content. Still, the intentional content of the conscious feeling is something like "I did A," while the content of the nonconscious process would be distinct. In many cases, the nonconscious content would have nothing to do with A. The con-

scious, intentional contents offered by Gazzaniga's split-brain patients hardly correspond to what appears to have been the actual cause of their behavior.[15] Consider too all the experiments in which nonconscious primes cause subjects to act in ways they do not consciously recognize or understand. In these sorts of cases, there is a radical mismatch between conscious and nonconscious contents. But even in cases where both contents represent the same action, their full intentional contents diverge nonetheless. In the content of the conscious feeling of willing—in (CF)—*I* am represented as the cause of my action, but that cannot be part of the content of nonconscious processes, since it is implausible that I, the agent, am represented at that level of processing.

3. The third consideration is that the empirical will is distributed. The actual causal processes are susceptible to social, causal influences. The processes that cause us to act are driven not merely by nonconscious factors internal to the agent but also by factors external to the agent—including, for example, a range of social factors described by Ross and Nisbett (1991). This means that the empirical will sometimes operates at an even greater distance from the feeling of willing. The feeling is a feeling of my having caused my own action. The feeling is that *I*, by virtue of a conscious intention, caused my action, that the cause of my action is wholly internal to me. There is nothing in this feeling to even suggest that the action was caused by factors external to my own

conscious awareness. In consequence, if, as Wegner claims, the empirical will is distributed over external causal factors, then the feeling of authorship does not correctly correspond, but rather conflicts, with the empirical will.[16]

These three considerations—that the feeling of conscious willing is false, that the intentional content (if any) of the empirical will often diverges from that of the conscious will, and that the empirical will is determined by factors external to the agent—all support the general claim that there is no meaningful sense in which the feeling of willing can correspond "correctly" with the empirical will.

But suppose I am wrong. Suppose there is some sort of correct correspondence between the feeling of will and the empirical will. Even so—this is my third point—such correspondence would be too weak to underwrite Wegner's claim about control and, in consequence, too weak to underwrite ascriptions of moral responsibility. The reason is this: even if it were true that the feeling of willing often corresponds correctly with the empirical will, that is not enough to justify the claim that the agent, in performing a given action, knowingly exerted the right kind of control. To justifiably claim that the agent was in control and hence deserves praise or blame, it must be the case that the agent can identify the action as her own. The agent must have good evidence that her action emanated from features of her empirical will that she recognizes as belonging to her character, as in some sense constituting who she is. But, as Wegner so delightfully describes, such evidence is not available from the first-person perspective, for there exists a host of conditions under which we sincerely but erroneously take ourselves to be the authors of our actions. A human agent cannot justifiably claim to know, merely on the basis of how things feel, that her action emanated in the right way from her character. The theoretical and experimental work to which Wegner appeals in the first eight chapters of his book constitutes, as I have been putting it, an epistemic defeater of the claim that one's action emanated in the relevant way from one's empirical will. We simply do not know our selves to that degree; the structure of our psychology works against it.[17]

My conclusion concerning the concepts 'control' and 'responsibility' is threefold. First, we should reject the claim that the feeling of willing serves the function of telling us when we have exercised control and when we have not. That claim is plausible only if we accept the correspondence claim and, as I have just argued, the correspondence claim is not plausible if we accept the theory of apparent mental causation. We should also reject the claim that the feeling of willing underwrites attributions of responsibility, on the grounds that the correspondence claim, even if true, is too weak. The conclusion to draw is that if we accept apparent mental causation we should not afford antecedent authority to either concept in framing our inquiries. That is the force of the directive in (DP).

We should insist that, with respect to these concepts, we first try to analyze inward and synthesize laterally. If that succeeds, the concepts will be vindicated. If it fails, the concepts will die or be transformed into something else. That is the force of (EC).

Second, it would indeed be useful to know the conditions under which the feeling of willing is a reliable guide to actual authorship. But if the theory of apparent mental causation is correct, and if (DP) and (EC) are on the right track, then the only way to discover that the feeling of will ever corresponds correctly with the actual causes of our actions is to resist the apparently natural tendency to trust that our feeling matches the actual causes of our action and instead achieve some distance from that feeling. We must try to discover whether and under what conditions the feeling correlates positively with nonconscious processes that cause us to act. Then we should try to discover whether the conscious feeling ever causes the nonconscious processes that cause our actions. To accomplish all this, however, we will have to discover the low-level, nonconscious processes that actually cause our actions.

Third, suppose we eventually discover that the feeling of willing rarely corresponds positively with the causes of action. What should we conclude? One suggestion is that we try to conform our conceptual and affective capacities to better fit the world. Golfers must learn to associate a host of conscious feelings with the actual movements of the club and with the direction, distance, and spin of the ball. They do this by developing "grooves" in their nervous systems with hundreds of hours of repetition, with the result that certain introspectively accessible feelings become reliable indicators of the efficacy of the swing and even the resulting trajectory of the ball. Something similar is true of opera singers who must learn to associate certain indicators that they consciously feel while singing with the specific pitch, tone, and volume of their voice. Perhaps there is something analogous to be learned about the feeling of willing. Perhaps we can train to become self-knowing agents the way golfers train to be golfers or the way opera singers train to be opera singers. Perhaps we can learn to correlate certain feelings of authorship with specific low-level causes of our actions. Perhaps. But the theory of apparent mental causation should make us cautious; we should doubt that the feeling of willing locks onto the actual causes of our actions. And most importantly from the progressive point of view, we should relinquish these cautious doubts only if we discover, as the result of adhering to my directives for inquiry, that the feeling does, or can be trained to, correspond with the empirical will.

***

At times, then, some of our best naturalists fall prey to conceptual conservatism. It is easy to see why this happens, especially for concepts as apparently important as 'moral responsibility.' Wegner is hardly alone in this. Daniel Dennett is another prominent naturalist who also lapses into conservatism with respect to 'moral responsibility.' He says, for example, that he and Wegner are in agreement on the fundamental importance of preserving the reality of moral responsibility, despite the apparently illusory status of actual conscious authorship: "I appreciate that there are people who will insist that Wegner's title is just right: He *is* showing that conscious will is an illusion.[But] Wegner eventually softens the blow by arguing that conscious will may be an illusion, but responsible, moral action is quite real. And that is the bottom line for both of us" (Dennett 2003, 224).

The importance of the directive to withhold antecedent authority from concepts dubious by psychological role is indeed hard to overstate. We should not assume that framing our inquiries in terms of our best scientific knowledge is a sufficient antidote to the ills of conservatism. The deliberate, self-conscious application of my directives is also prescribed.

In the final paragraph of his book, Wegner says, "Our sense of being a conscious agent who does things comes at a cost of being technically wrong all the time. The feeling of doing is how it seems, not what it is—but *that is as it should be. All is well* because the illusion makes us human" (Wegner 2002, 342; my italics). But is it really the illusion that makes us human—as opposed to the determination to live without illusions? And to what view of human nature ought we appeal to answer this question? We have, after all, good evidence that dubious concepts persist in our current conceptualization of human nature. At any rate, the attitude of exploration that drove Darwin and Wallace should incline us—as it inclined them—toward the construction of new concepts rather than the preservation of old ones. And we may, in the course of such explorations, run up against the limits of our nature—limits on our abilities as practical agents or our abilities as inquiring agents—but we should not pretend to know them in advance. Nor should we ignore what is indeed known already, namely, that the concept 'moral responsibility' is deeply dubious by descent and psychological role and, in consequence, bound to lead us in the wrong direction.

Conclusion

When compared with Kane's Romantic libertarianism, compatibilism appears far less extravagant and far closer to the truth. Upon inspection, however, we find an unhappy intellectual kinship. The problems afflicting libertarianism also afflict the compatibilist conceit. The types of philosophical theories canvassed in this chapter fail because they are framed by a view of our capacities as agents that has lost its former authority. Contemporary compatibilists are far too busy trying to resolve a conflict that has been superseded by the growth in human knowledge. The problem of understanding our capacities as agents in light of recent discoveries in psychology and neuroscience is largely ignored.

The problem with Wegner's view of responsibility is different. He is, as I say, among our finest naturalists. His way of framing inquiry into the question of free will fits beautifully the directives for inquiry described in this book. He does not approach the question of free will with a view of human agency undermined by contemporary work (including his own work); he insists instead on analyzing inward in an evolutionary framework with due skepticism concerning traditional concepts. And he claims to have discovered an interpretive system that requires with a vengeance the directive in (DP). Problems arise only when he turns to the concept that is perhaps the hardest to let go. Wegner's suggestion that the interpretive system posited in the theory of apparent mental causation serves anticipatory, systemic functions that contribute substantially to social order is indeed plausible. It may even be the case that discovering the anticipatory systemic functions of the mechanisms involved in apparent mental causation will someday enable us to articulate a novel, or at least a quite different, notion of responsibility. But it is abundantly clear already that the mechanisms that implement apparent mental causation cannot save our traditional humanistic concept of moral responsibility for the very simple reason that they undermine it! We must resist the inclination to save our traditional notion and endeavor to create an alternative way of understanding our selves as social agents.

I conclude, therefore, that the thesis of part 3 is established. A view of human agency that fails to account for the apparent facts of human freedom—including our overweening confidence that sometimes we are unfettered centers of command and control—is a failed view. But the two most plausible views of human freedom are in conflict with what we are learning about our actual capacities as agents. That is the upshot of the last two chapters. The conclusion, therefore, is that we do not know what kind of agents we are. We do not know what kind of freedom, if any, we have. This is not to deny what is surely obvious, namely, that we have very strong convictions and beliefs concerning human freedom and responsibility. But we know that the authority of these convictions and beliefs must be withheld precisely because they are so stubborn. Seeing this, we may conclude with confidence that we do not know in any substantive sense what kind of agents we are.

***

Darwin was indeed a shrewd rhetorician. He sensed the frailty of our historical and psychological imaginations and took steps to neutralize their ill effects. Throughout the *Origin* he draws attention to culturally inherited habits of thought, including our refusal to sum up the effects of small changes accumulated over vast stretches of time. He also draws attention to cognitive and affective dispositions that lead us astray, including our refusal to engrain in our minds the essential destructiveness of all life. Both refusals diminish our efforts at inquiry; both incline us toward conservatism and imperialism regarding dubious concepts. Future progress in knowledge, especially in areas plagued by dubious concepts, requires that we too take steps to neutralize the ill effects of our own frailties.

That is what this book has tried to do. The general response to the biases and infirmities that thwart our efforts at inquiry is the progressive orientation. This, of course, is a page out of Darwin, who clearly thought of naturalists as daring explorers charting unfamiliar territories. The more specific response is the directives for inquiry. These are constraints and guides keyed to some of our better-known weaknesses. They are tools that, having proven fruitful over the last couple centuries in the study of life, are perhaps likely to bear further knowledge in current and future studies of the human self. They are, at any rate, our most promising bet. Or so it seems to me.

The bet may be lost. Perhaps our best theories of the biases and infirmities that dog us are mistaken, or perhaps my understanding of those theories is. And perhaps my directives are wide of the mark and fail to effectively ameliorate our weaknesses. It is a very good bet, at any rate, that at least some of my directives will be rendered obsolete as our knowledge grows. That too is part of the progressive orientation. But the one claim on which we can safely bet the whole house is that, like it or not, we are subjects of the world. We are earthly (and earthy!) evolved animals descended from earlier but now extinct forms of life. So we must wonder: what are the prospects for knowing ourselves, for understanding what we really are, if we fail to subject our selves to the natural world in which we evolved into being?

Appendix

A Compilation of Directives for Naturalistic Inquiry

DESCRIPTIVE ACCURACY (DA)

When inquiring about any phenomena, identify the target of our investigation as fully as, but not more fully than, our initial grasp of the phenomena allows (chapter 4).

THEORETICAL COMPETITION (TC)

On the basis of the description of the phenomena that best adheres to (DA), develop alternative and competing theories of the phenomena and devise experiments to discover which theories

THE EXPECTATION OF CONCEPTUAL CHANGE (EC)

For systems we understand poorly or not at all, expect that, as inquiry progresses—as we analyze inward and synthesize laterally—the concepts in terms of which we conceptualize high-level systemic capacities will be altered or eliminated (chapter 3).

CONCEPTS DUBIOUS BY DESCENT (DD)

For any concept dubious by virtue of descent, do not make it a condition of adequacy on our philosophical theorizing that we preserve or otherwise "save" that concept; rather, bracket the concept with the expectation that it will be explained away or vindicated as inquiry progresses—as we analyze inward and synthesize laterally (chapter 3).

CONCEPTS DUBIOUS BY PSYCHOLOGICAL ROLE (DP)

For any concept dubious by psychological role, do not make it a condition of adequacy on our philosophical theorizing that we preserve or otherwise "save" that concept; rather, require that we identify the conditions (if any) under which the concept is correctly applied and withhold antecedent authority from that concept under all other conditions (chapter 3).

EVOLUTIONARY HISTORY (EH)

For any hypothesis regarding any human capacity, make it a condition of adequacy that, as we analyze inward and synthesize laterally, we do so within a framework informed by relevant considerations of our evolutionary history (chapter 3).

ANTICIPATORY SYSTEMIC FUNCTION (A)

For any psychological capacity of any minded organism, expect that among its most prominent systemic functions is the function of anticipating some feature of the organism's environment (chapter 3).

CONCEPT LOCATION PROJECT (CL)

For any concept dubious by descent, expect that the concept location project will fail; expect, that is, that the dubious elements of the traditional concept will face revision or elimination as we analyze inward and synthesize across the concepts and claims of all the relevant contemporary sciences (chapter 5).

NONCONSCIOUS MECHANISMS (NC)

For any conscious capacity of mind, expect that we will correctly understand that capacity only if we (1) frame our inquiry with plausible assumptions concerning our evolutionary history, (2) formulate competing hypotheses concerning the affective or cognitive capacities involved, and (3) analyze inward and synthesize laterally until we discover low-level, non-conscious, anticipatory mechanisms implementing the hypothesized capacities (chapter 6).

Notes

CHAPTER ONE

[1] These passages from Humboldt's *Cosmos* are quoted in Richards 2002 (520–21).

[2] Naive realism, as developed by Ross and Ward (1996) and more recently by Pronin, Lin, and Ross (2002), is discussed in chapter 7.

[3] A sustained and persuasive rebuttal to Richards's view of Darwin is offered by Ruse (2004).

[4] Similar discoveries were made by Etienne Geoffroy Saint-Hilaire and Georges Cuvier in France—an engrossing story of scientific discovery beautifully recounted by Appel (1987).

[5] Granting undue authority to vivid imaginings is the crux of the "irregular" argument from design in Hume's *Dialogues concerning Natural Religion.* The challenge from Hume's deist makes this clear: "Consider, anatomize the eye: Survey its structure and contrivance; and tell me, *from your own feeling,* if the idea of a contriver does not immediately flow in upon you with *a force like that of sensation*" (Hume 1779, 56; my italics).

CHAPTER TWO

[1] As we will see in chapter 4, Ruse (2003) illustrates the commitment to the first orientation.

[2] The relevant psychological theories are discussed beginning in chapter 6.

[3] Wegner's theory is explicated in chapter 7.

[4] As the discussion in chapters 6 and 7 attests.

CHAPTER THREE

[1] In the interests of convenience, an appendix to the book compiles all the directives discussed throughout the book.

[2] I gesture toward this point in chapter 4 of Davies 2001. Bechtel (2007, chapter 6) develops a compelling notion of *active* mechanisms to help explain salient features of living things.

[3] I harp on this point in chapter 6 of Davies 2001 and again in part 2 and part 3 of this book.

[4] More about our emotional registers in chapter 6.

[5] Michotte (1954) discusses several experiments in which subjects apply the concept 'cause' in the absence of relevant causal relations. Sperber, Premack, and Premack (1995) offer recent perspectives on the same theme. Malle, Moses, and Baldwin (2001) discuss our capacity to apply the concepts 'desire,' 'belief,' 'intention,' and the like. And, as mentioned, Wegner (2002) argues that the capacities with which we conceive of ourselves as free agents are powerful generators of agential illusions.

[6] Wolpert (1992) also discusses the unnaturalness of science.

[7] Humphreys (2004) suggests that technology already provides such devices. The epistemology

of science, he says, is no longer restricted to the epistemology of human beings.

[8] According to Lewin (1998), the term *hominine* has replaced the earlier term *hominid,* which was used to refer to the species in the human family or clade.

[9] The excellent discussion in Richardson 2007 is complemented by that in Buller 2005.

[10] This example is drawn from Panksepp 1998, chap.10 (see especially p.198). My discussion, like his, is intended to illustrate the importance of elementary evolutionary considerations in formulating our hypotheses concerning organismic capacities and, indeed, in settling on the correct description of those capacities.

[11] *Cycles of Contingency* is the title of an excellent book of essays (both pro and con) on the decentralized view of life to which I am here alluding (Oyama, Griffiths, and Gray 2001). I return to the decentralized view of life in chapter 5.

[12] For a bracing attack on nearly all contemporary metaphysics and a vigorous defense of a deeply naturalistic approach toward metaphysical theorizing, see Ladyman et al.2007. The progressive orientation defended in my discussion has much in common with the naturalism defended in theirs.

CHAPTER FOUR

[1] Readers who doubt this charge of murder—readers sympathetic to the idea that living things are genuinely designed, thanks to the mindless sorting and shaping of evolution by natural selection—are referred to chapter 5. That is where I argue that appeal to natural selection as a mindless designer is yet another form of conceptual conservatism.

[2] As mentioned, Ruse's view also flouts the directive in (DP). See chapter 6.

[3] There are educated doubts that complex traits always or even typically qualify as adaptations. Kauffman (1993) and others argue that complex, self-organized systems can and do emerge without natural selection—as, indeed, the materials upon which selection then acts. Lynch (2007) argues that the evolution of organismic complexity is the result, in no small part, of biological processes that are decidedly nonadaptive, including biases in mutation and gene conversion that have contributed to genome composition.

[4] And that belief appears dead wrong. As Broad (2005) reports, a team of investigators at the Harvard School of Dental Medicine uncovered compelling evidence that the tusk, far from being a weapon or ornament, is actually an extraordinary sense organ capable of detecting

changes in temperature, pressure, salinity gradients, and more.

[5] The competitive exclusion principle (also known as the Lotka-Volterra principle) asserts, roughly, that when two species occupy the same niche and compete for similar resources, it is highly probable that one will be superior to the other and eventually drive it to extinction. This principle will be satisfied, at any rate, unless the competing species are ecologically differentiated in some way.

[6] See Paley 1802 for one of the most elaborate statements of the view that Darwin so decisively killed.

[7] Incidentally, the strategies that motivate the expectation of conceptual change make clear why some recent attempts to challenge naturalism are wide of the mark. Perhaps widest of the mark is Alvin Plantinga's (2002) assertion that naturalism is defined as the denial of theism. Of course Plantinga is free to define naturalism as he wishes, but he should note that the progressive orientation defended in this book is a constraint on methodology, not a first-order claim about the composition of the universe. The expectation in (EC) makes no reference to any form of divinity. He should also note that the progressive orientation does not rule out theology by fiat. Were we to find that we can analyze into natural systems on the basis of

theological concepts and were we to discover low-level mechanisms that implement theological properties and converge on the taxonomies in other well-developed theories, then we would be required by our progressive orientation to engage the relevant theology. But that is not how things are at present. And nothing in Plantinga's views provides the slightest grounds for optimism.

CHAPTER FIVE

[1] I suspect that Ruse's second premise is true, which, as I argue in chapter 6, is grounds for concluding that the concepts 'function' and 'design' are dubious by psychological role.

[2] Enc (1979) and Schlosser (1998) are admirably clear on this point.

[3] One way to accomplish this is by positing "mechanisms" that are not the passive devices of our seventeenth-and eighteenth-century predecessors but that are rather active and systemic devices. In a marvelous discussion of this issue, William Bechtel (2007) describes conditions under which mechanisms comprising autonomous systems are active in a compelling way. Such mechanisms are, in any event, antithetical to Kant's nonmechanical formative power.

[4] Other advocates of proper functions (or near relatives to proper functions) include Allen and Bekoff (1995), Ayala (1970), Brandon (1981,

1990), Buller (1998), Enc (1979), Godfrey-Smith (1994), Griffiths (1993), Kitcher (1993), Neander (1991), Papineau (1993), Post (1991, 2006), Preston (1998), Price (1995), and Wright (1976). Proper functions have been thought useful for solving other philosophical problems. The following theories of linguistic or mental representation are, well, representative: Lycan 1988, McGinn 1989, Millikan 1984, Neander 1995, Sterelny 1990. See, too, the fine essays in MacDonald and Papineau 2006.

[5] The specific argument given in the next few pages is repeated from chapter 5 of Davies 2001. What is new is the broader context provided by the directives for inquiry.

[6] I am indebted to Don Ross for pressing this point and for the reference to Morris's book.

[7] Neander (1991) defends this claim.

[8] This point is well made by Amundson and Lauder (1994).

[9] A similar slide occurs in chapter 6 of Bechtel 2007. Building upon the concepts 'active mechanisms' and 'autonomous systems' (see note 3 above), Bechtel suggests we conceptualize living organisms as adaptive autonomous agents. Although this latter concept strikes me as theoretically rich and important, I do not agree that it provides us a naturalized sense of teleology. That would be plausible only if survival and reproduction qualified as genuine ends

of living organisms. But for reasons given in the above paragraph and in light of the directives already discussed (namely, [EC] and [DP]) and those discussed below ([DP] and [EH]), the attribution of suchends conflicts with a progressive orientation. The full argument for this claim, however, will have to be developed elsewhere.

[10] This is so, moreover, even if we grant that natural selection is the exclusive engine of evolutionary change—which, of course, it is not.

[11] This is true even if, as Post reports, Brandon himself embraces the existence of normative functions. Brandon's theory of normative functions is critically evaluated in Davies 2001, chap.6.

[12] And, as is clear in Robert Richards's 2002 discussion of the German Romantic naturalists, Kant's immediate successors were quick to appreciate the centrality of his presumed formative power, though they were far less restrained than Kant in their ontological pretensions. Indeed, the claim in Schelling, Goethe, and Humboldt is not that we must see living things *as if* animated by an intrinsic, form-imposing force but that living things literally *are* so animated.

[13] Though Oyama (2000) contests the use of the concept 'information.'

[14] Oyama, Griffiths, and Gray (2001) have a fine collection of essays on this issue.

[15] See Michael Ghislen 2003 for the same point aimed against Dawkins's account of selfish genes.

[16] *Evolutionary Naturalism* is the title of one of Ruse's many books.

[17] There is also the point that the causes of evolution are not restricted to selection. It is an empirical question how far selection (as opposed to other causes) has shaped the systemic functions of any trait. Proper functions appear ubiquitous only if we blithely assume that most systemic functions must be the effects of ancestral selection. But, as Bill Bechtel reminds me, this is not true. See Lynch 2007 for a recent discussion of nonadaptive sources of organismic complexity.

CHAPTER SIX

[1] See the discussion below of the second set of studies that support the theory of mind theory.

[2] On the relevance of pretend play to the theory of mind theory, see Leslie 1987.

[3] See also Leslie 1994 and 1995 for a developmental view with similarities to Tomasello's.

[4] See the studies in Moore and Dunham 1995 and the review in Carpenter, Nagell, and Tomasello 1998.

[5] One doubt that applies to Kelemen and Keil equally concerns the design of experiments that

depend on the maturing linguistic capacities of four-and five-year-olds. A doubt that applies to Kelemen's theory concerns the precise concept of 'function' or 'purpose' employed. In personal communication, Kelemen indicated that she has in mind the concept of normative functions that I have been criticizing throughout part 2. There is, however, the concept of systemic functions that I described in chapter 4 and defend in Davies 2001, and it would be interesting to know which concept of functions (if either) five-year-olds actually have in mind. Regarding Keil's view, we should wonder whether his experiments establish what he says they do. After all, a child may refuse to attribute knowledge to the disease-causing agent and still conceptualize the agent as minded, so long as the child thinks of the agent as appropriately ignorant or stupid. We should also wonder whether the fact that a child sometimes distinguishes between 'purpose' and 'knowledge' is enough to show that the conceptualizing capacities involved are distinct. And so on.

[6] I press this point in chapter 7.

[7] See Michael Gazzaniga's 1997 (chapter 3) discussion of Romi Nijhawan 1997.

[8] See Conway and Pleydell-Pearce 2000 and Conway 2003.

[9] Recall in this regard the point (discussed in chapter 3) from Norris and Inglehart (2004)

that our basic dispositions toward perceived threats appear to be calibrated early in life.

[10] See chapter 2 for discussion of the default assumption of all theologians and most philosophers.

[11] Not unlike Tomasello's intentional agents (discussed earlier in the chapter).

[12] Damasio focuses on the anticipatory function of certain affective states in practical reasoning. By contrast, Grush (2004) hypothesizes that various representational states involved in motor control also serve a crucial anticipatory function. The directive in (A), therefore, is more than an educated guess inspired by (EH). It is earning its explanatory keep in multiple areas of inquiry.

CHAPTER SEVEN

[1] More recent work on naive realism includes Pronin et al.2002 and Pronin, Gilovich, and Ross 2004. Pronin 2007 provides a brief and very clear overview of the biasing effects of naive realism and summarizes studies more recent than the ones discussed here.

[2] Pronin, Lin, and Ross (2002) aptly describe this as a *bias blind spot.*

[3] Additional troubling effects of naive realism include our stubborn ineptness in resolving interpersonal and intergroup conflicts. See Pronin, Puccio, and Ross 2002.

[4] There are, of course, faces that are attractive only when smiling warmly, just as there are attractive faces that are hard to take when expressing negative emotions. So the claim of invariance must be tempered. Still, I take it for granted that physical appearances are relatively invariant compared with warm and cold demeanors, and that seems enough to sustain the conclusion Nisbett and Wilson draw.

[5] Which means, of course, that had these judgments resulted in actions, the reasons given for those actions would have been demonstrably false.

[6] Pronin (2007) helpfully discusses what she calls the *introspection illusion.*

[7] Chapter 9 of Nisbett and Ross 1980 provides a clear overview of experiments and theories concerning our causal interpretive system.

[8] The theory described in the previous section may be mistaken. It is certainly not nearly as specific a theory as we would like. That, however, provides no relief to those who would resist the conclusions I have drawn thus far. We should have no patience for an alternative account of how human beings act on the basis of reasons unless the alternative is founded on a better-developed and more defensible theory of the actual mechanisms involved.

[9] It is no accident that the epigraph to the first chapter of Wegner's book is the quote from

Johnson that also serves as epigraph to this chapter.

[10] For experiments measuring the effects of such concepts on motor behavior, see Bargh, Chen, and Burrows 1996. For experiments on memory performance, see Dijksterhuis, Bargh, and Miedema 2000 and Dijksterhuis et al.2000. The discussion by Dijksterhuis and Bargh (2001) offers a helpful overview.

[11] Only somewhat puzzling because, as suggested by Glymour (2004), it is plausible to hypothesize that feelings of authorship contribute to various forms of learning.

[12] I critically assess Wegner's suggestions regarding control and moral responsibility in chapter 9.

CHAPTER EIGHT

[1] It is simply not plausible that all amplified neural processes affect all other neural processes. That would be to adopt the thesis that none of the processes of the brain are modular in their operations. The experiments in Goodale et al.1991 suggest otherwise.

[2] Advocates of agent causation sometimes announce that, according to their view, agents are "ontological primitives" and thus resistant to further analysis, as if that alleviated the mystery.

[3] It is plausible to construe "action" broadly to include "making a choice," in which case Wegn-

er's theory applies to the production of a choice or decision. The felt sense of effort in choosing to do A is the visceral effect of a causal inference, produced by the system that takes as inputs the prior thought of choosing A and the subsequent perception of one's self now intending to do A.

CHAPTER NINE

[1] Some philosophers have argued that free will and moral responsibility are possible *only if* determinism is true. That is to embrace the control conceit with open arms.

[2] An alternative version of the control conceit has been developed by Harry Frankfurt in a series of papers, including several in Frankfurt 1988.

[3] Fischer and Ravizza appeal to Peter Strawson's 1962 discussion of our natural reactive attitudes. And I suspect they will answer the above criticism of their view by claiming that the maturation process in which persons "take responsibility" is authentic in so far as it involves the maturation of our reactive attitudes. That is an effective reply only if Strawson's view is plausible. As I proceed to argue in the next section, however, Strawson's view is not plausible.

[4] As we will see, the claim that our reactive attitudes and attributions of responsibility are autonomous and hence insulated from theoretical

considerations echoes Wittgenstein's 1979 remarks on religion.

[5] See Strawson 1982, originally published in 1962. See Hume's *An Enquiry concerning Human Understanding* (1748).

[6] Nothing in Strawson's discussion suggests that our attributions of moral blame or praise correlate in a one-to-one fashion with the reactive attitudes. But this simplification helps illustrate what I take to be the strategy for locating 'moral responsibility' amid 'resentment' and 'gratitude.'

[7] It makes no difference which version of our traditional notion we settle on. If it contains any of the normative elements insisted on in our theological tradition, we must divest it of antecedent authority. Bernard Williams (1985) aptly describes our traditional notion of morality as "peculiar" and fleshes some of the elements that make it so. He also suggests an alternative notion of responsibility that is decidedly non-moral and less peculiar, what he calls "ethical" responsibility, descended to us from our pre-Christian heritage. I agree there is room for alternative notions of responsibility, but we should not settle upon any such notion unless led to it by inquiry informed by the directives described here.

[8] Strawson exegesis is no easy matter. It is possible to resist the interpretation just given. It is possible, in particular, to ignore or dismiss his

overt overtures to libertarians and insist that he is indeed eliminating our traditional (and mostly libertarian) notion of moral responsibility. I do not think that is a plausible reading of any ordinary language philosopher as subtle as Strawson. But if that is your preferred reading, you may skip ahead to my criticism of the second thesis of the forms of life conceit.

[9] The problem here is parallel to the problem discussed in chapter 5. The attempt by advocates of proper functions to locate our concept of normative functions in the mechanisms of evolution by natural selection leaves us empty-handed. The same is true of Strawson's attempt to locate 'moral responsibility' in our reactive attitudes.

[10] Panksepp emphasizes that his way of framing the inquiry into our capacities for social attachments may well evolve as we learn more. This is our current best bet.

[11] I wish to repeat the point that Strawson might be read as an eliminativist about 'moral responsibility.' That is, we might construe Strawson as identifying attributions of moral responsibility with the effects of our reactive attitudes. But in that case, there is only one thing, not two. And since the one thing on which Strawson insists the most is the centrality of our reactive attitudes, he must be eliminating the moral. This, as I say, is hard to square with

Strawson's overtures to libertarians. But even if it is his view, it fails to dodge my criticisms of the second thesis of the forms of life conceit.

[12] In which case, it cannot serve as a defense of the first thesis. That is, we might read Strawson as defending the proposed location of 'moral responsibility' in our reactive attitudes by insisting on the second thesis, by insisting that the connection between our attitudes and our attributions of moral responsibility is entrenched in an unalterable part of our psychology. But if the second thesis is indeed less defensible than the first, then this interpretation cannot save him.

[13] Of course, they might be the products of surgery. Science fiction writers might attempt to tell such stories, but on what grounds would we judge their stories as realistic, as grounded in the actual facts of our constitution?

[14] The remainder of this section comes from Davies 2007.

[15] One experimental subject, after seeing the command "take a walk" in her left field of vision only (thus receiving the information in the right hemisphere only), stood up and walked toward the door. When asked, "Why are you doing that?" the subject sincerely replied, "I need to get a drink."

[16] In pressing these last two points, I make no appeal to anything akin to a Cartesian Theater

(as described in Dennett 1991). The conscious feeling of will *may* be described with the sentence "*I* authored my *own* action," but it may also be described with the sentence "*this* bit of internal processing caused *that* bit of behavior." The second sentence assumes a capacity to take one's own internal processes as objects of thought or feeling. That requires no theater, unless the metaphor is stretched so thinly that every instance of having a thought or feeling somehow requires a spectator viewing a spectacle. And the first sentence may be permissible on the grounds that we do not yet know which bits of internal processing we ought to designate, in which case the second sentence asserts more than we can justifiably claim to know at present. (Thanks to Don Ross for pressing this point. Wegner's thoughts on this issue appear in his 2004 author's response to open peer commentary.)

[17] The Accuracy and the Knowledge assumptions described earlier in this chapter are relevant here.

Appreciation and Acknowledgments

The views defended in these chapters began taking shape five or six years ago, especially during the 2003–4 academic year when my research was funded by a National Endowment for the Humanities (NEH) fellowship. I am deeply grateful to the NEH for its generous support of my work. While still in rudimentary form, various parts were presented to audiences at James Madison University, University of Cincinnati, and the Southern Society for Philosophy and Psychology. I am grateful for the undeserved forbearance of so many good thinkers. As the ideas took shape they benefited from frequent, substantive conversations with George Harris and from his unflagging enthusiasm for the larger project. More recently and despite being gravely ill, George mustered the energy and concentration to give me valuable feedback on the final chapter. I am also indebted to Corina Vaida for conversations about the larger view and specific suggestions regarding the final chapter.

In the spring of 2005, the second meeting of the Mind and World Working Group was held at the University of Birmingham, Alabama. Thanks to the efforts of Harold Kincaid, Don Ross, and David Spurrett, that conference was superbly conceptualized and executed. In my presentation, I applied some of the elements of my progressive naturalism to Daniel Wegner's provocative theory of apparent mental causation and especially to the view of moral responsibility with which he concludes his book. I received stimulating questions and helpful encouragement from several thinkers, including George Ainslie, Don Ross, Philip Pettit, Daniel Dennett, and Daniel Wegner.

After that conference, Harold Kincaid, Don Ross, and David Spurrett, with help from Lynn Stephens, produced an excellent anthology of essays titled *Distributed Cognition and the Will: Individual Volition and Social Context.* My contribution to that volume, "What Kind of Agent Are We? A Naturalistic Framework for the Study of Human Agency," is reproduced in part 3 of this book, distributed over parts of chapters 7 and 9. I am grateful to MIT Press for permission to use that material in the wider context of this discussion.

The manuscript readers for the University of Chicago Press were Bill Bechtel and Don Ross. I know this because, at my request, both sent me work of their own that helped clarify and sharpen the views defended here. I am deeply indebted to Bill and Don not only for constructive criticisms and suggestions but also for invigorating encouragement regarding the larger project. Their words of support helped quiet some of the doubts that incessantly rang in my ears while writing the final chapters. Readers of this book should not fail to study the works by Bill and Don cited herein.

A few weeks before the book was scheduled to go into production, I wrote to Emily Pronin, a psychologist at Princeton University whom I had never met, and asked for her criticisms of chapter 7. As I had hoped, she corrected some of my descriptions of published experiments, highlighted points not altogether in focus, and referred me to additional literature. Emily does important work on naive realism—see chapter 7 for references—and in addition she is a generous colleague who, across intellectual disciplines, extends herself even to a stranger.

It is a particular pleasure to thank Elizabeth Branch Dyson, editor at the University of Chicago Press. From my earliest contact with the Press, Elizabeth has been a wholehearted advocate of the book, always animated with grace and good cheer, ready with suggestions for improving the quality of the book and strategies for increasing its readership. Some of her offerings were the fruits of consultation with other editors at the Press—a press, evidently, that believes in editors who work together to cross disciplinary boundaries, editors who do not let their authors hide behind stultifying academic jargon, editors who think that books ought to be intellectual and aesthetic wholes. The concerted efforts of Elizabeth and her colleagues have been an abiding source of encouragement for me and should be an inspiration for serious writers and readers.

This book, not merely the effort of producing it but also the heart of the views espoused, is dedicated to my daughter Cassie. The limits and infirmities of the human imagination get us in trouble when we are trying to acquire knowledge. Or so I think. In other areas of life, a feeble imagination, even the will to enfeeble one's imagination, helps us avoid various sorts of trouble. Most every father and mother knows that becoming a parent creates vulnerabilities that can crush the imagination. Certain thoughts become unthinkable, or at least unbearable. And in my own case, being the father to such startling intelligence and beauty, and to a convulsive excitement for the world that her growing body just barely contains, is, at times, simply too much. An infirm imagination, thank goodness, protects me from what I know.

Love sometimes enfeebles the imagination and pushes knowledge away; love and knowledge do not always get along. We are emotional and cognitive aggregates cast from a merciless history that cares nothing for integration. Our constitutional conflicts even include the naive conviction that we are free of such conflicts. That is the kind of agent we are. The only consolation—if that is what this is—is that at certain moments, if we are lucky, we get to know and understand a few truths about the world, and then, at other moments, if we are lucky, our imaginations go to sleep and we, as fathers and mothers, get lost in the convulsing intelligence and beauty of new and ascending life.

References

Allen, Colin, and Marc Bekoff. 1995. Biological function, adaptation, and natural design. *Philosophy of Science* 62:609–22.

Amundson, Ron, and George Lauder. 1994. Function without purposes: The uses of causal role function in evolutionary biology. *Biology and Philosophy* 9:443–69.

Appel, Toby. 1987. *The Cuvier-Geoffroy debate: French biology in the decades before Darwin.* Oxford: Oxford University Press.

Ayala, Francisco. 1970. Teleological explanations in evolutionary biology. *Philosophy of Science* 37:1–15.

Bargh, John, Mark Chen, and Lara Burrows. 1996. Automaticity of social behavior: Direct effects of trait construct and stereotype activation on action. *Journal of Personality and Social Psychology* 71:230–44.

Bargh, John, Peter Gollwitzer, A. Lee-Chai, K. Barndollar, and R. Troetschel. 2001. The automated will: Nonconscious activation and pursuit of behavioral goals. *Journal of Personality and Social Psychology* 81:1014–27.

Bechtel, William. 2007. *Mental mechanisms: Philosophical perspectives on cognitive neuroscience.* Hillsdale, NJ: Lawrence Erlbaum.

Bechtel, William, and Robert Richardson. 1993. *Discovering complexity: Decomposition and localization as strategies in scientific research.* Princeton, NJ: Princeton University Press.

Bergson, Henri. 1907. *Creative evolution.* Trans. Arthur Mitchell [1911]. New York: Holt.

Blumenbach, Johann Friedrich. 1781. *Uberden Bildungstrieb* [On the formative power]. Gottingen: Johann Christian Dieterich.

Boogerd, F.C., F.J. Bruggeman, R.C. Richardson, A. Stephan, and H.V. Westerhoff. 2005. Emergence and its place in nature: A case study of biochemical networks. *Synthese* 145:131–64.

Boswell, James. 1791. *Life of Johnson.* New York: Oxford University Press.

Brandon, Robert. 1981. Biological teleology: Questions and explanations. *Studies in the History and Philosophy of Science* 12:91–105.

______. 1990. *Adaptation and environment.* Princeton, NJ: Princeton University Press.

Broad, William. 2005. It's sensitive. Really. *New York Times,* December 13. http://www.nytimes.com/2005/12/13/science/13narw.html?scp=1&sq=It%27s+Sensitive.+Really.&st=nyt.

Buller, David. 1998. Etiological theories of function: A geographical survey. *Biology and Philosophy* 13:505–27.

______. 2005. *Adapting minds: Evolutionary psychology and the persistent quest for human nature.* Cambridge, MA: MIT Press.

Byrne, Richard, and Andrew Whiten. 1988. *Machiavellian intelligence: Social expertise and the evolution of intellect in monkeys, apes, and humans.* Oxford: Oxford University Press.

Carpenter, M., K. Nagell, and Michael Tomasello. 1998. Social cognition, joint attention, and communicative competence from 9 to 15 months of age. *Monographs for the Society for Research in Child Development* 63.

Cheney, Dorothy, and Robert Seyfarth. 2007. *Baboon metaphysics: The evolution of a social mind.* Chicago: University of Chicago Press.

Chisholm, Roderick. 1964. Human freedom and the self. The Lindley Lecture, University of Kansas.

Conway, Martin. 2003. Cognitive-affective mechanisms and processes in autobiographical memory. *Memory* 11, no.2:217–24.

Conway, Martin, and Christopher Pleydell-Pearce. 2000. The construction of autobiographical memories in the memory system. *Psychological Review* 107, no.2:261–88.

Cummins, Robert. 1975. Functional analysis. *Journal of Philos* ophy 72:741–60.

______. 1977. Programs in the explanation of behavior. *Philosophy of Science* 44:269–87.

______. 1983. *The Nature of psychological explanation.* Cambridge, MA: MIT Press.

Damasio, Antonio. 1994. *Descartes' error: Emotion, reason, and the human brain.* New York: G.P. Putnam's Sons.

Darwin, Charles. 1859. *On the origin of species by means of natural selection, or the preservation of favoured races in the struggle for life.* A facsimile of the first edition with an introduction by Ernst Mayr, 1964. Cambridge, MA: Harvard University Press.

______. 1871. *The descent of man.* A photo reproduction by Princeton University Press, 1981. Princeton, NJ.

______. 1872a. *The expression of the emotions in man and animals.* Introduction, afterword, and commentaries by Paul Ekman. Oxford: Oxford University Press, 1998.

______. 1872b. *The origin of species by means of natural selection, or the preservation of favoured races in the struggle for life.* 6th ed. London, John Murray.

______. 1969. *The autobiography of Charles Darwin, 1809–1882.* Ed. Nora Barlow. New York: W.W. Norton.

Davies, Paul Sheldon. 2000. The nature of natural norms: Why selected functions are systemic capacity functions. *Noûs* 34:85–107.

______. 2001. *Norms of nature: Naturalism and the nature of functions.* Cambridge, MA: MIT Press.

______. 2007. What kind of agent are we? A naturalistic framework for the study of human agency. In *Distributed cognition and the will: Individual volition and social context,* ed. Don Ross,

David Spurrett, Harold Kincaid, and Lynn Stephens. Cambridge, MA: MIT Press.

Dawkins, Richard. 1976. *The selfish gene.* Oxford: Oxford University Press.

Dennett, Daniel. 1991. *Consciousness explained.* Boston: Little, Brown, and Company.

_____. 2003. *Freedom evolves.* New York: Penguin Books.

Descartes, Rene. 1641 [1985]. *Meditations on first philosophy.* In *The philosophical writings of Descartes vol. II.* Trans. John Cottingham, Robert Stoothoff, and Dugald Murdoch. Cambridge: Cambridge University Press.

Dijksterhuis Ap, H. Aarts, John Bargh, and A. van Knippenberg. 2000. On the relation between associative strength and automatic behavior. *Journal of Experimental Social Psychology* 36:531–44.

Dijksterhuis, Ap, and John Bargh. 2001. The perception-behavior expressway: Automatic effects of social perception on social behavior. In vol.33 of *Advances in experimental social psychology,* ed. M.P. Zanna, 1–40. San Diego: Academic Press.

Dijksterhuis Ap, John Bargh, and J. Miedema. 2000. Of men and mackerels: Attention and automatic behavior. In *Subjective experience in social cognition and behavior,* ed. H. Bless and J.P. Forgas, 36–51. Philadelphia: Psychology Press.

Dreifus, Claudia. 2007. Small wonders: Understanding the way of the warrior sperm. *New York Times,* January 23. http://www.nytimes.com/2007/01/23/science/23conv.html?scp=1&sq=Small+Wonders%3A+Understanding+the+Way+of+the+Warrior+Sperm&st=nyt.

Dretske, Fred. 1995. *Naturalizing the mind.* Cambridge, MA: MIT Press.

Enc, Berent. 1979. Function attributions and functional explanations. *Philosophy of Science* 46:343–65.

Evans, E.M. 2001. Cognitive and contextual factors in the emergence of diverse belief systems: Creation versus evolution. *Cognitive Psychology* 42:217–66.

Fischer, John Martin, and Mark Ravizza. 1998. *Responsibility and control: A theory of moral responsibility.* New York: Cambridge University Press.

Fitzsimmons, Gráinne, and John Bargh. 2003. Thinking of you: Nonconscious pursuit of interpersonal goals associated with relationship partners. *Journal of Personality and Social Psychology* 83:148–64.

Frankfurt, Harry. 1988. *The importance of what we care about: Philosophical essays.* New York: Cambridge University Press.

Gazzaniga, Michael. 1997. *The mind's past.* Berkeley: University of California Press.

Gazzaniga, Michael, and Joseph LeDoux. 1978. *The integrated mind.* New York: Plenum.

Ghislen, Michael. 2003. Selfish chromosomal deletions and other delusions for which bioeconomics provides viable alternatives. *Journal of Interdisciplinary Economics* 14:373–79.

Gilovich, Thomas. 1990. Differential construal and the false consensus effect. *Journal of Personality and Social Psychology* 59:623–34.

Glymour, Clark. 2004. We believe in freedom of the will so that we can learn. *Behavioral and Brain Sciences* 27, no.5:661–62.

Godfrey-Smith, Peter. 1994. A modern history theory of functions. *Noûs* 28:344–62.

Goethals, George, and Richard Reckman. 1973. The perception of consistency in attitudes. *Journal of Experimental Social Psychology* 9:491–501.

Goodale, Melvin, A. Milner, L. Jakobsen, and D. Carey. 1991. Perceiving the world and grasping it: A neurological dissociation. *Nature* 349:154–56.

Griffiths, Paul. 1993. Functional analysis and proper functions. *British Journal for the Philosophy of Science* 44:409–22.

Grush, Rick. 2004. The emulation theory of representation: Motor control, imagery, and perception. *Behavioral and Brain Sciences* 27:377–442.

Hobson, J. Allan. 1994. *Dreaming as delirium: How the brain goes out of its mind.* Cambridge, MA: MIT Press.

Humboldt, Alexander von. 1858. *Cosmos: A sketch of the physical description of the universe.* Vol.1. Trans. E.C. Otte. Baltimore: Johns Hopkins University Press, 1997. [Reprint of the English-language edition published by Harper and Brothers in 1858.]

Hume, David. 1739–40. *A treatise of human nature.* Ed. L.A. Selby-Bigge. Oxford: Oxford University Press. [1st ed. 1888.]

_____. 1748 [1975]. *An enquiry concerning human understanding.* In *Enquiries concerning human understanding and concerning the principles of morals.* Ed. L.A. Selby-Bigge and P.H. Nidditch. Oxford: Oxford University Press.

_____. 1779 [1980]. *Dialogues concerning natural religion.* Ed. Richard Popkin. Indianapolis: Hackett.

Humphrey, Nicholas. 1986. *The inner eye: Social intelligence in evolution.* Oxford: Oxford University Press.

Humphreys, Paul. 2004. *Extending ourselves: Computational science, empiricism, and scientific method,* New York: Oxford University Press.

Jablonka, Eva, and Marion Lamb. 2005. *Evolution in four dimensions: Genetic, epigenetic, behavioral, and symbolic variation in the history of life.* Cambridge, MA: MIT Press.

Jackson, Frank. 1998. *From metaphysics to ethics: A defence of conceptual analysis.* Oxford: Oxford University Press.

Kane, Robert. 1996. *The significance of free will.* New York: Oxford University Press.

Kanner, L. 1943. Autistic disturbance of affective conduct. *Nervous Child* 2:217–50.

Kant, Immanuel. 1790. *Critique of judgment.* Trans. J.H. Bernard, 1951. New York: Hafner Press.

Kauffman, Stuart. 1993. *The origins of order: Self-organization and selection in evolution.* New York: Oxford University Press.

Keil, Frank. 1994. The birth and nurturance of concepts by domains: The origins of concepts of living things. In *Mapping the mind: Domain specificity in cognition and culture,* ed. Lawrence Hirschfeld and Susan Gelman, 234–54. Cambridge: Cambridge University Press.

Kelemen, Deborah. 1999. The scope of teleological thinking in preschool children. *Cognition* 70:214–72.

______. 2004. Are children "intuitive theists"? Reasoning about purpose and design in nature. *Psychological Science* 15:295–301.

Kitcher, Philip. 1993. Function and design. *Midwest Studies in Philosophy* SVIII:379–97.

Ladyman, James, Don Ross, David Spurrett, and John Collier. 2007. *Everything must go: metaphysics naturalized.* New York: Oxford University Press.

Leslie, Alan. 1987. Pretense and representation: The origins of "theory of mind." *Psychological Review* 94:412–26.

_____. 1994. ToMM, ToBy, and agency: Core architecture and domain specificity. In *Mapping the mind: Domain specificity in cognition and culture,* ed.

Lawrence Hirschfeld and Susan Gelman, 119–48. Cambridge: Cambridge University Press.

_____. 1995. A theory of agency. In *Causal cognition: A multidisciplinary debate,* ed. Dan Sperber, David Premack, and Ann James Premack, 121–41. Oxford: Oxford University Press.

_____. 2000. "Theory of mind" as a mechanism of selective attention. In *The New Cognitive Neurosciences,* 2nd ed., ed. Michael Gazzaniga, 1235–47. Cambridge, MA: MIT Press.

Leslie, Alan, and L. Thaiss. 1992. Domain specificity in conceptual development: Neuropsychological evidence from autism. *Cognition* 43:225–51.

Levi-Setti, Riccardo. 1993. *Trilobites.* 2nd ed. Chicago: University of Chicago Press.

Lewin, Roger. 1998. *Principles of human evolution.* Malden, MA: Blackwell Science.

Lord, Charles, Lee Ross, and Mark Lepper. 1979. Biased assimilation and attitude polarization: The effects of prior theories on subsequently considered evidence. *Journal of Personality and Social Psychology* 37, no.11:2098–109.

Lycan, William. 1988. *Judgment and justification.* New York: Cambridge University Press.

Lynch, Michael. 2007. The frailty of adaptive hypotheses for the origins of organismal complexity. *Proceedings of the National Academy of Sciences* 104:8597–604.

MacDonald, Graham, and David Papineau, eds. 2006. *Teleosemantics.* Oxford: Oxford University Press.

Malle, Bertram, Louis Moses, and Dare Baldwin. 2001. *Intentions and intentionality: Foundations of social cognition.* Cambridge, MA: MIT Press.

McCauley, Robert. 2000. The naturalness of religion and the unnaturalness of science. In *Explanation and cognition,* ed. Frank Keil and Robert Wilson, 61–85. Cambridge, MA: MIT Press.

McGinn, Colin. 1989. *Mental content.* Oxford: Basil Blackwell.

Michotte, Albert. 1954 [1963]. *The perception of causality.* Trans. T.R. Miles. New York: Basic Books.

Millikan, Ruth. 1984. *Language, thought, and other biological categories.* Cambridge, MA: MIT Press.

_____. 1989. In defense of proper functions. *Philosophy of Science* 56:288–302.

Moore, C., and P. Dunham, eds. 1995. *Joint attention: Its origin and role in development.* Hillsdale, NJ: Erlbaum.

Morris, Conway. 2003. *Life's solutions: Inevitable humans in a lonely universe.* Cambridge: Cambridge University Press.

Neander, Karen. 1991. Functions as selected effects: The conceptual analyst's defense. *Philosophy of Science* 58:168–84.

_____. 1995. Misrepresenting and malfunctioning. *Philosophical Studies* 79:109–41.

Niebuhr, Reinhold. 1941. *The nature and destiny of man: A Christian interpretation.* Vol.1, *Human nature.* New York: Charles Scribner's Sons.

Nietzsche, Friedrich. 1974. *The gay science.* Trans. Walter Kaufmann. New York: Random House. [First published 1882.]

Nijhawan, Romi. 1997. Visual decomposition of color through motion extrapolation. *Nature* 386:66–69.

Nisbett, Richard, and Lee Ross. 1980. *Human inference: Strategies and shortcomings of social judgment.* Englewood Cliffs, NJ: Prentice Hall.

Nisbett, Richard, and Timothy Wilson. 1977a. The halo effect: Evidence for unconscious alteration of judgments. *Journal of Personality and Social Psychology* 35:250–56.

______. 1977b. Telling more than we can know: Verbal reports on mental processes. *Psychological Review* 84:231–59.

Norris, Pippa, and Ronald Inglehart. 2004. *Sacred and secular: Religion and politics worldwide.* New York: Cambridge University Press.

Oyama, Susan. 2000. *Evolution's eye: A systems view of the biology-culture divide.* Durham, NC: Duke University Press.

Oyama, Susan, Paul Griffiths, and Russell Gray. 2001. *Cycles of contingency: Developmental systems and evolution.* Cambridge, MA: MIT Press.

Paley, William. 1802. *Natural theology, or Evidences of the existence and attributes of the deity collected from the appearances of nature.* London: Farnborough, Gregg [reprint 1970].

Panksepp, Jaak. 1998. *Affective neuroscience.* New York: Oxford University Press.

Papineau, David. 1993. *Philosophical naturalism.* Oxford: Blackwell.

______. 2001. The status of teleosemantics, or How to stop worrying about swampman. *Australasian Journal of Philosophy* 79:279–89.

Plantinga, Alvin. 2002. The evolutionary argument against naturalism: An initial statement of the argument. In *Naturalism defeated? Essays on Plantinga's evolutionary argument against naturalism,* ed. James Beilby, 1–12. Ithaca, NY: Cornell University Press.

Post, John 1991. *Metaphysics: A contemporary introduction.* New York: Paragon House.

______. 2006. Naturalism, reduction, and normativity: Pressing from below. *Philosophy and Phenomenological Research* 1:1–27.

Premack, David. 1990. The infant's theory of self-propelled objects. *Cognition* 36, no.1:1–16.

Preston, Beth. 1998. Why is a wing like a spoon? A pluralist theory of functions. *Journal of Philosophy* 95:215–54.

Price, Carolyn. 1995. Functional explanations and natural norms. *Ratio* n.s.7:143–60.

Pronin, Emily. 2007. Perception and misperception of bias in human judgment. *Trends in Cognitive Science* 11:37–43.

Pronin, Emily, Thomas Gilovich, and Lee Ross. 2004. Objectivity in the eye of the beholder: Divergent perceptions of bias in self versus others. *Psychological Review* 111:781–99.

Pronin, Emily, Daniel Lin, and Lee Ross. 2002. The bias blind spot: Perceptions of bias in self versus others. *Personality and Social Psychology Bulletin* 28:369–81.

Pronin, Emily, Carolyn Puccio, and Lee Ross. 2002. Understanding misunderstanding: Social psychological perspectives. In *Heuristics and biases: The psychology of intuitive judgment,* ed. Thomas Gilovich, Dale Griffin, and Daniel Kahneman. New York: Cambridge University Press, 2002.

Richards, Robert. 2002. *The romantic conception of life: Science and philosophy in the age of Goethe.* Chicago: University of Chicago Press.

Richardson, Robert. 2007. *Evolutionary psychology as maladapted psychology.* Cambridge, MA: MIT Press.

Richerson, Peter, and Robert Boyd. 2005. *Not by genes alone: How culture transformed human evolution.* Chicago: University of Chicago Press.

Ross, Lee, David Greene, and Pamela House. 1977. The false consensus effect: An egocentric bias in social perception and attribution processes. *Journal of Experimental Social Psychology* 13:279–301.

Ross, Lee, and Richard Nisbett. 1991. *The person and the situation: Perspectives of social psychology.* Philadelphia: Temple University Press.

Ross, Lee, and A. Ward. 1996. Naïve realism in everyday life: Implications for social conflict and misunderstanding. In *Values and knowledge,* ed. T.

Brown, E. Reed, and E. Turiel, 103–35. Hillsdale, NJ: Erlbaum.

Ruse, Michael. 1995. *Evolutionary naturalism: Selected essays.* London: Routledge.

______. 2003. *Darwin and design: Does evolution have a purpose?* Cambridge, MA: Harvard University Press.

______. 2004. The romantic conception of Robert J. Richards. *Journal of the History of Biology* 37:3–23.

Schlosser, G. 1998. Self-re-production and functionality: A systems-theoretical approach to teleological explanation. *Synthese* 116:303–54.

Sperber, Dan, David Premack, and Ann James Premack. 1995. *Causal cognition: A multidisciplinary debate.* Oxford: Oxford University Press.

Sterelny, Kim. 1990. *The representational theory of mind.* Oxford: Blackwell.

Strawson, Peter. 1982. Freedom and resentment. In *Free Will,* ed. Gary Watson. Oxford: Oxford University Press. First published 1962 in *Proceedings of the British Academy* 48:1–25.

Tomasello, Michael. 1995. Joint attention as social cognition. In *Joint attention: Its origins and role in development.* Ed. C. Moore and P. Dunham, 103–30. Hillsdale, NJ: Erlbaum.

_____. 1999. *The cultural origins of human cognition.* Cambridge, MA: Harvard University Press.

Tomasello, Michael, and Josep Call. 1997. *Primate cognition.* New York: Oxford University Press.

Valins, Stuart, and Alice Ray. 1967. Effects of cognitive desensitization on avoidance behavior. *Journal of Personality and Social Psychology* 7:345–50.

Wegner, Daniel. 2002. *The illusion of conscious will.* Cambridge, MA: MIT Press.

_____. 2004. Frequently asked questions about conscious will. [Author's response to open peer commentary.] *Behavioral and Brain Sciences* 27, no.5:679–92.

_____. 2007. Self is magic. In *Psychology and Free Will,* ed. J. Baer, J.C. Kaufman, and R.F. Baumeister. New York: Oxford University Press.

Wegner, Daniel, Valerie Fuller, and Betsy Sparrow. 2004. Clever hands: Uncontrolled intelligence in facilitated communication. *Journal of Personality and Social Psychology* 85, no.1:5–19.

Wegner, Daniel, and Thalia Wheatley. 1999. Apparent mental causation: Sources of the experience of will. *American Psychologist* 54, no.7:480–92.

Whiten, Andrew, and Richard Byrne. 1997. *Machiavellian intelligence.* Vol.2, *Extensions and evaluations.* Cambridge: Cambridge University Press.

Williams, Bernard. 1985. *Ethics and the limits of philosophy.* Cambridge, MA: Harvard University Press.

Wilson, Timothy. 2002. *Strangers to ourselves.* Cambridge, MA: Harvard University Press.

Wilson, Timothy, P. Laser, and J. Stone. 1982. Judging the predictors of one's own mood: Accuracy and the use of shared theories. *Journal of Experimental Social Psychology* 18:537–56.

Wittgenstein, Ludwig von. 1979. *Remarks on Frazer's golden bough.* Ed. Rush Rhees. Trans. A.C. Miles. Atlantic Highlands, NJ: Humanities Press.

Wolpert, Lewis. 1992. *The unnatural nature of science.* Cambridge, MA: Harvard University Press.

Wright, Larry. 1976. *Teleological explanations: An etiological analysis of goals and functions.* Berkeley: University of California Press.

Zaitchik, Deborah. 1990. When representations conflict with reality: The preschooler's problem with false beliefs and "false" photographs. *Cognition* 35:41–68.

Zimmer, Carl. 2005. Plain, simple, primitive? Not the jellyfish. *New York Times,* June 21. http://www.nytimes.com/2005/06/21/science/21jell.html?_r=1&scp=1&sq=Plain%2C+simple%2C+primitive%3F+Not+the+jellyfish.&st=nyt&oref=slogin.

Front Cover Flap

Subjects of the World
DARWIN'S RHETORIC AND THE STUDY OF AGENCY IN NATURE
Paul Sheldon Davies

Being human while trying to scientifically study human nature confronts us with our most vexing problem. Efforts to explicate the human mind are thwarted by our cultural biases and entrenched infirmities; our first-person experiences as practical agents convince us that we have capacities beyond the reach of scientific explanation. What we need to move forward in our understanding of human agency, Paul Sheldon Davies argues, is a reform in the way we study ourselves and a long-overdue break with traditional humanist thinking.

Davies locates a model for change in the rhetorical strategies employed by Charles Darwin in *On the Origin of Species.* Darwin worked hard to anticipate and diminish the anxieties and biases that his radically historical view of life was bound to provoke. Likewise, Davies draws from the history of science and contemporary psychology and neuroscience to build a framework for the study of human agency that identifies and diminishes outdated and limiting biases. The result is a heady, philosophically wide-ranging argument in favor of recognizing that humans are, like everything else, subjects of the natural world—an acknowledgment that may free us to see the world the way it actually is.

Back Cover Flap

PAUL SHELDON DAVIES is the author of *Norms of Nature: Naturalism and the Nature of Functions.* He lives in Virginia with his wife and daughter and teaches philosophy at the College of William and Mary.

Back Cover Material

"This is a highly original and very provocative book. Davies puts forth a version of naturalism that is far more critical of our philosophical and intellectual heritage than past proponents have dared to be. Sharply and forcefully argued, it will be of interest to a substantial range of philosophers, biologists, cognitive scientists, and lay readers."

WILLIAM BECHTEL, UNIVERSITY OF CALIFORNIA, SAN DIEGO

"As a psychologist working at the edges of philosophy, I found this work clear, penetrating, and deliciously relevant to the scientific study of the problem of conscious will. Topics of human agency and the experience of being an agent have confused more than one thinker, but there is no confusion here. This book builds a sturdy bridge between the naturalistic philosophy of mind and the science of psychology that many readers will want to cross."

DANIEL M. WEGNER, HARVARD UNIVERSITY

"Paul Davies takes us on a logically and rhetorically compelling modern search for human agency. This outstanding analysis, well informed by naturalistic views of our evolved affective nature, is the kind of philosophical work that is essential for a field to move forward when ever-increasing findings from modern science are inconsistent with traditional philosophical arguments. This book is for all who wish to immerse themselves in the modern search for free will. It is steeped in the rich liqueur of current scientific and philosophical perspectives and delusions."

JAAK PANKSEPP, WASHINGTON STATE UNIVERSITY

A

B

C

D

E

F

G

H

I

J

K

O

P

T

V

W

Z